The Populist Moment

THE POPULIST
MOMENT

A Short History of the
Agrarian Revolt in America

LAWRENCE GOODWYN

OXFORD UNIVERSITY PRESS
Oxford   London   New York
1978

OXFORD UNIVERSITY PRESS

Oxford London Glasgow
New York Toronto Melbourne Wellington
Ibadan Nairobi Dar es Salaam Cape Town
Kuala Lumpur Singapore Jakarta Hong Kong Tokyo
Delhi Bombay Calcutta Madras Karachi

This book is an abridged edition of *DEMOCRATIC PROMISE:
The Populist Moment in America* (Oxford University Press, 1976)

Library of Congress Cataloging in Publication Data

Goodwyn, Lawrence.
The Populist moment.

Abridged ed. of Democratic promise.
Bibliography: p.
Includes index.
1. Populism—United States—History.
2. United States—Politics and government—
1865–1900. I. Title.
E669.G672 1978   239'.88'009   78-1349
ISBN 0-19-502416-8
ISBN 0-19-502417-6 pbk.

Printed in the United States of America

First published by Oxford University Press, 1978
First issued as an Oxford University Press paperback, 1978

To Nell

Introduction

This book is about the flowering of the largest democratic mass movement in American history. It is also necessarily a book about democracy itself. Finally it is about why Americans have far less democracy than they like to think and what would have to happen to alter that situation.

The passionate events that are the subject of this book had their origins in the social circumstances of a hundred years ago when the American population contained huge masses of farmers. A large number of people in the United States discovered that the economic premises of their society were working against them. These premises were reputed to be democratic—America after all was a democratic society in the eyes of most of its own citizens and in the eyes of the world—but farmers by the millions found that this claim was not supported by the events governing their lives.

The nation's agriculturalists had worried and grumbled about "the new rules of commerce" ever since the prosperity that accompanied the Civil War had turned into widespread distress soon after the war ended. During the 1870's they did the kinds of things that concerned people generally do in an effort to cope with "hard times." In an occupation noted for hard work they worked even harder. When this failed to change things millions of families migrated westward in an effort to enlist nature's help. They were driven by the thought that through sheer physical labor they might wring more production from the new virgin lands of the West than they had been able to do in their native states of Ohio and Virginia and Alabama. But, though railroad land agents created beguiling stories of Western prosperity, the men and women who listened, and went, found that the laws of commerce worked against them just as much in Kansas and Texas as they had back home on the eastern side of the Mississippi River.

So in the 1870's, the farmers increasingly talked to each other about their troubles and read books on economics in an effort to discover what had gone wrong. Some of them formed organizations of economic self-help like the Grange and others assisted in pioneering new institutions of political self-help like the Greenback Party. But as the hard times of the 1870's turned into the even harder times of the 1880's, it was clear that these efforts were not really going anywhere. Indeed, by 1888 it was evident that things were worse than they had been in 1878 or 1868. More and more people saw their farm mortgages foreclosed. As everyone in rural America knew, this statistic inexorably yielded another, more ominous one: the number of landless tenant farmers in America rose steadily year after year. Meanwhile, millions of small landowners hung on grimly, their unpaid debts thrusting them dangerously close to the brink of tenantry and peonage. Hard work availed nothing. Everywhere the explanation of events was the same: "Times were hard."

Then gradually, in certain specific ways and for certain specific reasons, American farmers developed new methods that enabled them to try to regain a measure of control over their own lives. Their efforts, halting and disjointed at first, gathered form and force until they grew into a coordinated mass movement that stretched across the American continent from the Atlantic coast to the Pacific. Millions of people came to believe fervently that a wholesale overhauling of their society was going to happen in their lifetimes. A democratic "new day" was coming to America. This whirlwind of effort, and the massive upsurge of democratic hopes that accompanied it, has come to be known as the Populist Revolt. This book is about that moment of historical time. It seeks to trace the planting, growth, and death of the mass democratic movement known as Populism.

For a number of reasons, all of them rather fundamental to historical analysis, the Populist moment has proved very difficult for Americans to understand. Under the circumstances, it is probably just as well to take these reasons up one at a time at the very outset in an effort to clear away as much underbrush as possible before turning our attention to the protesting farmers of the 1890's.

There are three principal areas of interpretive confusion that bear directly on the Populist experience. First, very little understanding exists as to just what mass democratic movements are, and how they happen. Second, there are serious problems embedded in the very language of description modern Americans routinely employ to characterize political events. These problems particularly affect commonly held presumptions about how certain "classes" of people are supposed to "act" on the stage of history. Finally, and by all odds most importantly, our greatest problem in understanding protest is grounded in contemporary American culture. In addition to being central, this cultural difficulty is also the most resistant to clear explanation: we are not only culturally confused, our confusion makes it difficult for us even to imagine our confusion. Obviously, it is prudent, then, to start here.

The reigning American presumption about the American experience is grounded in the idea of progress, the conviction that the present is "better" than the past and the future will bring still more betterment. This reassuring belief rests securely on statistical charts and tables certifying the steady upward tilt in economic production. Admittedly, social problems have persisted—inequities of income and opportunity have plagued the society—but these, too, have steadily been addressed through the sheer growth of the economy. For all of its shortcomings, the system works.

This is a powerful assumption. It may be tested by reflecting upon the fact that, despite American progress, the society has been forced to endure sundry movements of protest. In our effort to address the inconvenient topic of protest, our need to be intellectually consistent—while thinking within the framework of continuous progress—has produced a number of explanations about the nature of dissent in America. Closely followed, these arguments are not really explanations at all, but rather the assertion of more presumptions that have the effect of defending the basic intuition about progress itself. The most common of these explanations rests upon what is perceived to be a temporary malfunction of the economic order: people protest when "times are hard." When times stop being "hard," people stop protesting and things return to "normal"—that is to say, progress is resumed.*

Unfortunately, history does not support the notion that mass protest movements develop because of hard times. Depressed economies or exploitive arrangements of power and privilege may produce lean years or even lean lifetimes for millions of people, but the historical evidence is conclusive that they do not produce mass political insurgency. The simple fact of the matter is that, in ways that affect mind and body, times have been "hard" for most humans throughout human history and for most of that period people have not been in rebellion. Indeed, traditionalists in a number of societies have often pointed in glee to this passivity, choosing to call it "apathy" and citing it as a justification for maintaining things as they are.

This apparent absence of popular vigor is traceable, however, not to apathy but to the very raw materials of history—that complex of rules, manners, power relationships, and memories that collectively comprise what is called culture. "The masses" do not rebel in instinctive response to hard times and exploitation because they have been culturally organized by their societies not to rebel. They have, instead, been instructed in deference. Needless to say, this is the kind of social circumstance that is not readily apparent to the millions who live within it.

The lack of visible mass political activity on the part of modern

---

* Of course protest is not invariably an economic expression; it can also emerge from unsanctioned conceptions of civil liberty, as illustrated by the movements of Anti-Federalists, suffragettes, feminists, and blacks. While demonstrably important in their own terms, such movements historically have not mounted broad challenges to the underlying economic structures of inherited power and privilege that fundamentally shape the parameters of American society. Even the one movement that most nearly approached this level of insurgency—abolitionism—actually challenged, in slavery, only a deviance *within* the economic order rather than the underlying structure of the order itself.

industrial populations is a function of how these societies have been shaped by the various economic or political elites who fashioned them. In fundamental ways, this shaping process (which is now quite mature in America) bears directly not only upon our ability to grasp the meaning of American Populism, but our ability to understand protest generally and, most important of all, on our ability to comprehend the prerequisites for democracy itself. This shaping process, therefore, merits some attention.

Upon the consolidation of power, the first duty of revolutionaries (whether of the "bourgeois" or "proletarian" variety) is obviously to try to deflect any further revolutions that necessarily would be directed against them. Though a strong central police or army has sometimes proved essential to this stabilizing process, revolutionaries, like other humans, do not yearn to spend their lives fighting down counterrevolutions. A far more permanent and thus far more desirable solution to the task of achieving domestic tranquillity is cultural—the creation of mass modes of thought that literally make the need for major additional social changes difficult for the mass of the population to imagine. When and if achieved, these conforming modes of thought and conduct constitute the new culture itself. The ultimate victory is nailed into place, therefore, only when the population has been persuaded to define all conceivable political activity within the limits of existing custom. Such a society can genuinely be described as "stable." Thenceforth, protest will pose no ultimate threat because the protesters will necessarily conceive of their options as being so limited that even should they be successful, the resulting "reforms" will not alter significantly the inherited modes of power and privilege. Protest under such conditions of cultural narrowness is, therefore, not only permissible in the eyes of those who rule, but is, from time to time, positively desirable because it fortifies the popular understanding that the society is functioning "democratically." Though for millions of Americans the fact is beyond imagining, such cultural dynamics describe politics in contemporary America. It is one of the purposes of this book to trace how this happened.

It can be said, in advance of the evidence, that this condition of social constraint is by no means solely an American one; it is worldwide and traceable to a common source: the Industrial Revolution. Over the last eight generations, increasingly sophisticated systems of economic organization have developed throughout the western world, spawning factories and factory towns and new forms of corporate centralization and corporate politics. Through these generations of the modern era, millions have been levered off the land and into cities to provide the human components of the age of machinery. Meanwhile, ownership of both industrial and agricultural land has been increasingly centralized. Yet, though these events have caused massive dislocations of family, habitat, and work, creating mass suffering in many societies and anxiety in all of them, mass movements of protest have rarely materialized. This historical constant points to a deeper reality of the modern world: industrial societies have not only become centralized, they have devised rules of conduct that are intimidating to their populations as a whole. Though varying in intensity in important ways from nation to nation, this has now happened everywhere—whether a particular society regards itself as "socialist" or "capitalist." When people discover that their intellectual autonomy has become severely circumscribed and their creativity forcibly channeled into acceptable non-political modes of expression (a not unfrequent circumstance in socialist systems of economic organization), they are told that their autonomous hungers are "decadent," "individualistic," and, if obstinately pursued, will be seen as "revisionist" and "counterrevolutionary" in intent. On the other hand, when people discover they have far fewer opportunities than others of their countrymen (a not infrequent circumstance in capitalist systems of economic organization), they are told—as Populists were told in the 1890's and as blacks, Appalachian whites, and migrant laborers are told today in America—that they are "improvident," "lazy," inherently "deprived," or in some similar fashion culturally handicapped and at fault. These stigmas (which in earlier times were also visited upon Irish, Jewish, Italian, and other immigrants to America) generate fears; people are driven to undergo considerable indignity to earn sufficient status to avoid them. Accordingly, they try to do those things necessary to "get ahead." The result is visible in the obsequious day-to-day lives of white-collar corporate employees in America—and in the even more obsequious lives of Communist Party functionaries in the Soviet Union. Though life clearly contains far more options in America than in Russia, the persistence of these varying modes of mass deference in both countries illuminate the social limits of democratic forms in modern industrial societies generally. It is interesting to observe that each of the aforementioned adjectives, from "counterrevolutionary" to "lazy," is offered in the name of preserving corporate or state cultures self-described as "democratic." It is clear that the varied methods of social control fashioned in industrial societies have, over time, become sufficiently pervasive and subtle that a gradual erosion of democratic aspirations among whole populations has taken place. Accordingly, it is evident that the precise meaning of the word "democracy" has become increasingly obscure as industrialization has proceeded. It is appropriate to attempt to pursue the matter—for problems inherent in defining democracy underscore the cultural crisis of modern life around the globe.

In America, an important juncture in the political consolidation of the industrial culture came some four generations ago, at the culmination of the Populist moment in the 1890's. Because the decline in popular democratic aspiration since then has involved an absence of something rather than a visible presence, it has materialized in ways that are largely unseen. Politically, the form exists today primarily as a mass folkway of resignation, one that has become increasingly visible since the end of World War II. People do not believe they can do much "in politics" to affect substantively either their own daily lives or the inherited patterns of power and privilege within their society. Nothing illustrates the general truth of this phenomenon more than the most recent exception to it, namely the conduct of the student radicals of the 1960's. While the students themselves clearly felt they could substantively affect "inherited patterns of power and privilege," the prevailing judgment of the 1970's, shared by both the radicals and their conservative critics, is that the students were näive to have had such sweeping hopes. Today, political life in America has once more returned to normal levels of resignation.

Again, the folkway is scarcely an American monopoly. In diverse forms, popular resignation is visible from Illinois to the Ukraine. It does more than measure a sense of impotence among masses of people; it has engendered escapist modes of private conduct that focus upon material acquisition. The young of both societies seek to "plug in" to the system, the better to reap private rewards. Public life is much lower on the scale of priorities. Indeed, the disappearance of a visible public ethic and sense of commonweal has become the subject of hand-wringing editorials in publications as diverse as the Chicago *Tribune* in the United States and *Izvestia* in the Soviet Union. The retreat of the Russian populace represents a simple acknowledgment of ruthless state power. Deference is an essential ingredient of personal survival. In America, on the other hand, mass resignation represents a public manifestation of a private loss, a decline in what people think they have a political right to aspire to—in essence, a decline of individual political self-respect on the part of millions of people.

The principal hazard to a clear understanding of the meaning of American Populism exists in this central anomaly of contemporary American culture. Reform movements such as Populism necessarily call into question the underlying values of the larger society. But if that society is perceived by its members to be progressive and democratic—and yet is also known to have resisted the movement of democratic reform—the reigning cultural presumption necessarily induces people to place the "blame" for the failure of protest upon the protesters themselves. Accordingly, in the case of the Populists, the mainstream presumption is both simple and largely unconscious: one studies Populism to learn where the Populists went wrong. The condescension toward the past that is implicit in the idea of progress merely reinforces such complacent premises.

Further, if the population is politically resigned (believing the dogma of "democracy" on a superficial public level but not believing it privately) it becomes quite difficult for people to grasp the scope of popular hopes that were alive in an earlier time when democratic expectations were larger than those people permit themselves to have today. By conjoining these two contradictory features of modern culture—the assumption of economic progress with massive political resignation—it is at once evident that modern people are culturally programmed, as it were, to conclude that past American egalitarians such as the Populists were "foolish" to have had such large democratic hopes. Again, our "progressive" impulse to condescend to the past merely reinforces such a presumption. In a society in which sophisticated deference masks private resignation, the democratic dreams of the Populists have been difficult for twentieth-century people to imagine. Contemporary American culture itself therefore operates to obscure the Populist experience.

A second obstacle to a clear perception of Populism is embedded in the language of description through which contemporary Americans attempt to characterize "politics." A central interpretive tool, derived from Marx but almost universally employed today by Marxists and non-Marxists alike, is based upon concepts of class: that is, that the intricate nature of social interaction in history can be rendered more intelligible by an understanding of the mode and extent of class conflict that was or was not at work during a given period. Needless to say, many psychological, social, and economic ingredients are embedded in concepts of class, and, when handled with care, they can, indeed, bring considerable clarity to historical events of great complexity. Nevertheless, as an interpretive device, "class" is a treacherous tool if handled casually and routinely—as it frequently is. For example, offhand "class analysis," when applied to the agrarian revolt in America, will merely succeed in rendering the Populist experience invisible. While classes in agricultural societies contain various shadings of "property-consciousness" on the part of rich landowners, smallholders, and landless laborers ("gentry," "farmers," and "tenants," in American terminology), these distinctions create more problems than they solve when applied to the agrarian revolt. It is a long-standing assumption—not so thoroughly tested in America by sustained historical investigation as some might believe—that "landowners" must perforce behave in politically reactionary ways. The political aspirations of the landless are seen to deserve intense scrutiny, but the politics of "the landed" cannot be expected to contain serious progressive ideas. The power of this theoretical assumption can scarcely be understated. It permits the political efforts of millions of human beings to be dismissed with the casual flourish of an abstract category of interpretation. One can only assert the conviction that a thoroughgoing history of, for example, the Socialist Party of the United States, including the history of the recruitment of its agrarian following in early twentieth-century America, will not be fully pieced together until this category of political analysis is successfully transcended. The condtiion of being "landed" or "landless" does not, à priori, predetermine one's potential for "progressive" political action: circumstances surrounding the ownership or non-ownership of land are centrally relevant, too. The Populist experience in any case puts this proposition to a direct and precise test, for the agrarian movement was created by landed and landless people. The platform of the movement argued in behalf of the landless because that platform was seen as being progressive for small landowners, too. Indeed, from beginning to end, the chief Populist theoreticians—"landowners" all—stood in economic terms with the propertyless rural and urban people of America.

In consequence, neither the human experiences within the mass institutions generated by the agrarian revolt nor the ideology of Populism itself can be expected to become readily discernible to anyone, capitalist or Marxist, who is easily consoled by the presumed analytical clarity of categories of class. The interior life of the agrarian revolt makes this clear enough.* While the economic and political threads of populism did not always mesh in easy harmony (any more than the cultural threads did), the evolution of the political ideology of the movement proceeded from a common center and a common

* Though European and Asian conceptions of agricultural "classes" can be applied to America only if one is willing to accept a considerable distortion of reality, Populism can with a stretch of the imagination be seen as a product of the organizing efforts of middle peasants engaged in recruiting both their own "kind" and lower peasants. But one must immediately add that such interesting examples of agrarian "unity" can be more swiftly explained through recourse either to the labor theory of value or to simple historical observation rather than to class categories.

experience and thus possessed an instructive degree of sequential consistency.†

The use of the word "sequential" provides an appropriate introduction to the final hazard confronting the student of the agrarian revolt—the rather elementary problem of defining just what "mass movements" are and how they happen. The sober fact is that movements of mass democratic protest—that is to say, coordinated insurgent actions by hundreds of thousands or millions of people—represent a political, an organizational, and above all, a cultural achievement of the first magnitude. Beyond this, mass protest requires a high order not only of cultural education and tactical achievement, it requires a high order of *sequential* achievement. These evolving stages of achievement are essential if large numbers of intimidated people are to generate both the psychological autonomy and the practical means to challenge culturally sanctioned authority. A failure at any stage of the sequential process aborts or at the very least sharply limits the growth of the popular movement. Unfortunately, the overwhelming nature of the impediments to these stages of sequential achievement are rarely taken into account. The simple fact of the matter is that so difficult has the process of movement-building proven to be since the onset of industrialization in the western world that all democratic protest movements have been aborted or limited in this manner prior to the recruitment of their full natural constituency. The underlying social reality is, therefore, one that is not generally kept firmly in mind as an operative dynamic of modern society—namely, that mass democratic movements are overarchingly difficult for human beings to generate.

How does mass protest happen at all, then—to the extent that it does happen?

The Populist revolt—the most elaborate example of mass insurgency we have in American history—provides an abundance of evidence that can be applied in answering this question.

† For example, five sequentially related stages of this ideological process, all contradicting conclusions implicit in perfunctory class analysis, are treated on pp. 75–76, 78–80, 84–87, 91–93 and 108–13.

The sequential process of democratic movement-building will
be seen to involve four stages: (1) the creation of an autonomous
institution where new interpretations can materialize that run
counter to those of prevailing authority—a development which,
for the sake of simplicity, we may describe as "the movement
forming"; (2) the creation of a tactical means to attract masses
of people—"the movement recruiting"; (3) the achievement of
a heretofore culturally unsanctioned level of social analysis—
"the movement educating"; and (4) the creation of an institutional means whereby the new ideas, shared now by the rank
and file of the mass movement, can be expressed in an autonomous political way—"the movement politicized."

Imposing cultural roadblocks stand in the way of a democratic
movement at every stage of this sequential process, causing
losses in the potential constituencies that are to be incorporated
into the movement. Many people may not be successfully
"recruited," many who are recruited may not become adequately
"educated," and many who are educated may fail the final test
of moving into autonomous political action. The forces of
orthodoxy, occupying the most culturally sanctioned command
posts in the society, can be counted upon, out of self-interest,
to oppose each stage of the sequential process—particularly the
latter stages, when the threat posed by the movement has become
clear to all. In the aggregate, the struggle to create a mass
democratic movement involves intense cultural conflict with
many built-in advantages accruing to the partisans of the established order.

Offered here in broad outline, then, is a conceptual framework
through which to view the building process of mass democratic
movements in modern industrial societies. The recruiting, educating, and politicizing methods will naturally vary from movement to movement and from nation to nation, and the relative
success in each stage will obviously vary also.* The actions of

* American factory workers, for example, were unable for generations to
successfully complete step one of the process because their initial strikes for
recognition were lost and their fragile new unions destroyed. They thus were
unable to create autonomous institutions of their own. See pp. 41–42, 117–18,
and 174–76.

both the insurgents and the defenders of the received culture can also be counted upon to influence events dramatically.

Within this broad framework, it seems helpful to specify certain subsidiary components. Democratic movements are initiated by people who have individually managed to attain a high level of personal political self-respect. They are not resigned; they are not intimidated. To put it another way, they are not culturally organized to conform to established hierarchical forms. Their sense of autonomy permits them to dare to try to change things by seeking to influence others. The subsequent stages of recruitment and of internal economic and political education (steps two, three, and four) turn on the ability of the democratic organizers to develop widespread methods of internal communication within the mass movement. Such democratic facilities provide the only way the movement can defend itself to its own adherents in the face of the adverse interpretations certain to emanate from the received culture. If the movement is able to achieve this level of internal communication and democracy, and the ranks accordingly grow in numbers and in political consciousness, a new plateau of social possibility comes within reach of all participants. In intellectual terms, the generating force of this new mass mode of behavior may be rather simply described as "a new way of looking at things." It constitutes a new and heretofore unsanctioned mass folkway of autonomy. In psychological terms, its appearance reflects the development within the movement of a new kind of collective self-confidence. "Individual self-respect" and "collective self-confidence" constitute, then, the cultural building blocks of mass democratic politics. Their development permits people to conceive of the idea of acting in self-generated democratic ways—as distinct from passively participating in various hierarchical modes bequeathed by the received culture. In this study of Populism, I have given a name to this plateau of cooperative and democratic conduct. I have called it "the movement culture." Once attained, it opens up new vistas of social possibility, vistas that are less clouded by inherited assumptions. I suggest that all significant mass democratic movements in human history have generated this autonomous capacity. Indeed, had they not done so, one cannot visualize how they could have developed into significant mass democratic movements.*

Democratic politics hinge fundamentally on these sequential relationships. Yet, quite obviously the process is extremely difficult for human beings to set in motion and even more difficult to maintain—a fact that helps explain why genuinely democratic cultures have not yet been developed by mankind. Self-evidently, mass democratic societies cannot be created until the components of the creating process have been theoretically delineated and have subsequently come to be understood in practical ways by masses of people. This level of political analysis has not yet been reached, despite the theoretical labors of Adam Smith, Karl Marx, and their sundry disciples and critics. As a necessary consequence, twentieth-century people, instead of participating in democratic cultures, live in hierarchical cultures, "capitalist" and "socialist," that merely call themselves democratic.

All of the foregoing constitutes an attempt to clear enough cultural and ideological landscape to permit an unhampered view of American Populism. The development of the democratic movement was sequential. The organizational base of the agrarian revolt was an institution called the National Farmers Alliance and Industrial Union. Created by men of discernible self-possession and political self-respect, the Alliance experimented in new methods of economic self-help. After nine years of trial and error, the people of the Alliance developed a powerful mechanism of mass recruitment—the world's first large-scale working class cooperative. Farmers by the hundreds of thousands

---

* The political terminology offered in this study is meant to be inclusive rather than exclusive. The terms "movement forming," "movement recruiting," "movement educating," and "movement politicized," plus the sub-categories of "individual self-respect," "collective self-confidence," and "internal communications," together with the summary phrase, "movement culture," all embrace a certain measure of abstraction. More precise terminology would be helpful and clearly needs to be developed. On the other hand, the spacious Marxist abstraction, "class consciousness," is simply too grand to have precise meaning. Though the term was pathbreaking when first developed, it is too unwieldy to describe human actions with the kind of specificity needed to make sense of the hierarchical hazards and democratic opportunities existing within complex twentieth-century social systems.

flocked into the Alliance. In its recruiting phase, the movement swept through whole states "like a cyclone" because, easily enough, the farmers joined the Alliance in order to join the Alliance cooperative. The subsequent experiences of millions of farmers within their cooperatives proceeded to "educate" them about the prevailing forms of economic power and privilege in America. This process of education was further elaborated through a far-flung agency of internal communication, the 40,000 lecturers of the Alliance lecturing system. Finally, after the effort of the Alliance at economic self-help had been defeated by the financial and political institutions of industrial America, the people of the movement turned to independent political action by creating their own institution, the People's Party. All of these experiences, stretching over a fifteen-year period from 1877 to 1892, may be seen as an evolutionary pattern of democratic organizing activity that generated, and in turn was generated by, an increasing self-awareness on the part of the participants. In consequence, a mass democratic movement was fashioned.*

Once established in 1892, the People's Party challenged the corporate state and the creed of progress it put forward. It challenged, in sum, the world we live in today. Though our loyalty to our own world makes the agrarian revolt culturally difficult to grasp, Populism may nevertheless be seen as a time of economically coherent democratic striving. Having said this, it is also necessary to add that Populists were not supernatural beings. As theoreticians concerned with certain forms of capitalist exploitation, they were creative and, in a number of ways, prescient. As economists, they were considerably more thoughtful and practical than their contemporary political rivals in both major parties. As organizers of a huge democratic movement, Populists learned a great deal about both the power of the

* * Since Populism was a mass movement (and one that attempted to be even more "massive" than it was), the sequential stages in the recruiting and politicizing of its mass constituency are the core of this study. The stages of this sequential development may be found on pp. 26–35, 39–41, 49, 58–59, 64–66, 73, 75–87, 91–93, 108–15, 125–36, 148–64, and 172–82. The mass movement reached its practical range of politicization in 1892 (pp. 175–82). The summary interpretation of these sequential democratic stages is on pp. 293–310 and 318–19.

received hierarchy and the demands imposed on themselves by independent political action. As third party tacticians, they had their moments, though most of their successes came earlier in the political phase of their movement than later. And, finally, as participants in the democratic creed, they were, on the evidence, far more advanced than most Americans, then or since.

But American Populists did not parachute in from Mars. They grew up in American culture and felt the pull of its teachings. Though they knew they were pioneers, and earnestly endeavored to persuade others of the merit of what they had learned along their own path of democratic innovation, they did not always do so free of inherited cultural barnacles. They had earlier learned a number of things taught by the dominant culture; and more than a few people stumbled into the movement with many of their traditional inheritances almost wholly intact. The tension between these modes of conduct persisted within the agrarian movement throughout its life. Populists also encountered more specific hazards. They sought to enlist the urban working class without understanding the needs, nor the barriers to autonomous political expression, that informed life in the metropolitan ghettos of the nineteenth-century factory worker. Populists sought to enlist landless black sharecroppers (and in so doing explored new modes of interracial political coalition) without ultimately shaking off the more subtle forms of white supremacy that fundamentally undermine the civility of American society in our own time. And Populists tried through democratic politics to bring the corporate state under popular control without fully anticipating the counter-tactics available to the nation's financial and industrial spokesmen. In summary, though Populists generated a vibrant democratic movement, they were not unfailingly guided by genius. Their shortcomings as well as their achievements contain much that is useful to all those who study history because they continue to nurse aspirations for an industrial society in which generous social relations among masses of people might prevail as a cultural norm.

A final prefatory comment. It is helpful to bear in mind that the Populist moment in America came before the global twentieth-century struggle between the East and the West. It came, therefore, at a time when the range of culturally sanctioned political traditions was broader than two. As children of the two spreading cultures of intellectual conformity that are a product of that conflict, modern people live in a time of extreme politicization of knowledge throughout the world. Rigid modes of thought and terminology dominate the schools and colleges of both traditions. The young of both can imbibe the particular received wisdom of their theoretical tradition (however distorted by the events of history that theoretical tradition has become) or, if they are somehow unconvinced and can cope with the ostracism involved, they can adopt the rival mode. Within the perceived limits of this most ideologically confined of recent centuries, one is surely right: man is either a competitive being or a cooperative being. However all those who are not persuaded by this speculation—or faith—soon discover how difficult it is to express their disbelief in terminology that the confined participants in twentieth-century culture can understand. Capitalist "modernization theory" and Marxist "democratic centralism," together with supporting linguistic accoutrements, have left mankind in our time with few conceptual options through which to assert believable political aspirations to the mass of the world's peoples. In both traditions, one "believes" or one does not, but in terms of sanctioned categories of political language, the option for the unconvinced is an option of one. So be it. The Populists did not know that the Russian Revolution, the Chinese Revolution, and the ascendancy of the multi-national corporation were to be the coercive and competitive products of the industrial age. Spared the ideological apologetics and narrowness of a later time, Populists thought of man as being both competitive and cooperative. They tilted strongly toward the latter, but they also confronted the enduring qualities of the former. They accepted this complexity about mankind, and they tried to conceive of a society that would be generous—and would also house this complexity. With all of their shortcomings, including theoretical shortcomings, the Populists speak to the anxieties of the twentieth century with their own unique brand of rustic relevance.

Out of their cooperative struggle came a new democratic

community. It engendered within millions of people what Martin Luther King would later call a "sense of somebodiness." This "sense" was a new way of thinking about oneself and about democracy. Thus armed, the Populists attempted to insulate themselves against being intimidated by the enormous political, economic, and social pressures that accompanied the emergence of corporate America.

To describe that attempt is to describe their movement.

L. G.

Durham, N.C.
May 1, 1978

Acknowledgments

To the various acknowledgments specifically pertaining to Populist research that I noted in the unabridged edition of this book, I wish to take this opportunity to add two of a less specific nature. Both relate to those sundry and elusive intellectual influences that shape one's personal perspective.

I doubt that one can overstate the impact that *Origins of the New South* by C. Vann Woodward had on those students of American history who reached maturity in the 1950's. Surrounded as we were by historical tomes of limp apology for the Bourbon past, *Origins* opened up new ways of thinking about both the Southern and the national heritage. All historians whose work touches on post-Reconstruction America have ever since stood on Woodward's shoulders. This work is no exception.

The most creative democratic theorist that I have been privileged to know on a long and intimate basis is my good friend Harry Boyte. Alert to the ways people can fashion a politics supportive of mass democratic aspirations, he has been alert also to the ways the agrarian crusade encouraged impoverished people to "see themselves" experimenting in democratic forms. The sentence to that effect that forms part of the frontispiece of this study grew out of conversations between us over this manuscript and reflects our mutual respect for another historian, E. P. Thompson. In less specific but even more helpful ways, Harry Boyte has deepened my understanding of the democratic impulse as an embattled but central component of the human experience.

L.G.

Contents

I hope we shall crush in its birth the aristocracy of our monied corporations which dare already to challenge our government to a trial of strength, and bid defiance to the laws of our country.

THOMAS JEFFERSON, 1816

The people need to "see themselves" experimenting in democratic forms.

I

Creating a Democratic Politics

He was the largest landholder ... in one county and Justice
of the Peace in the next and election commissioner in both,
and hence the fountainhead if not of law at least of advice and
suggestion. ... He was a farmer, a usurer, a veterinarian;
Judge Benbow of Jefferson once said of him that a milder-
mannered man never bled a mule or stuffed a ballot box. He
owned most of the good land in the county and held mortgages
on most of the rest. He owned the store and the cotton gin
and the combined grist mill and blacksmith shop in the village
proper and it was considered, to put it mildly, bad luck for a
man of the neighborhood to do his trading or gin his cotton
or grind his meal or shoe his stock anywhere else.

The furnishing merchant in *The Hamlet*, by William Faulkner

The suballiance is a schoolroom.

Alliance lecturer

1

Prelude to Populism
Discovering the Limits of American Politics

"The people are near-sighted. . . ."

People understood that the war—the Civil War—had changed everything in America. For four bloody years, the war itself had dominated the life of the nation, and when peace came, the memory of the war shaped the way people acted. Almost a generation after the guns had fallen silent, the nation's new poet of democracy, Walt Whitman, was saying that the essence of history, philosophy, art, poetry, and even personal character "for all future America" would trace its sources back to the war.

The flamboyant poet left one thing out. The war also revolutionized the way Americans thought about politics. After Appomattox, the nation's party system had become so fundamentally altered that the change indeed seemed to be "for all future time." The war, not political ideas, dominated the new American party system.

To a number of thoughtful Americans, the crucial postwar topic for the nation—as important in the long run as the future status of the freedman—concerned the need to reorganize the country's exploitive banking system to bring a measure of economic fairness to the "plain people," white as well as black. However this idea—like almost any economic idea in post-Civil War America—confronted national political constituencies seemingly impervious to new concepts of any kind. This state of affairs was traceable to a party system which had been so massively altered by the war that the new situation seemed to make Whitman's sweeping description appear to be an under-statement.

The old Jacksonian resonances of Whig-Democratic conflict, containing as they did still older rhythms of the Jeffersonian-Federalist struggle, were all but obliterated by the massive realignment of party constituencies that had accompanied the war and its aftermath. The memories and even some of the slogans of ancestral debates still persisted in the postwar American ethos, but they no longer possessed a secure political home. Sectional, religious, and racial loyalties and prejudices were used to organize the nation's two major parties into vast coalitions that ignored the economic interests of millions.

The post-Civil War nation contained three basic occupational groups. Ranked in order of numbers, they were farmers, urban workers, and the commercial classes. The war had divided the three groups into six constituencies—Northern farmers, work-ers, and men of commerce, and their counterparts in the South. Two additional groups were defined less by occupation than by caste—free Negroes in the North and ex-slaves in the South—making a total of eight broad classifications. It was a striking feature of post-Civil War politics that a substantial majority of persons within seven of these eight constituencies followed their wartime sympathies, "voting as they shot," into the 1890's.

The sole exception was the urban working class of the North, which fought for the Union but voted in heavy majorities for the rebel-tainted Democratic Party. For the voters in this class, sectional loyalty had given way before religious and racial loyalties as the prime determinants of political affiliation. Uneas-ily adrift in a sea of Yankee Protestant Republicanism, the largely immigrant and overwhelmingly Catholic urban workers clustered defensively in makeshift political lifeboats fashioned after the Tammany model. Generally run by Irish bosses, these scattered municipal vessels essentially conveyed patronage and protection. Though nominally Democratic, little in their design reflected Jeffersonian patterns; their chief function, aside from affording their captains a measure of income and status, was to provide the immigrant masses with local security in an alien world. The Catholic Democratic tendency, defensive though it was, encouraged a reaction among Protestants that they themselves considered defensive. By inescapable, if circular logic, it provided many thousands of Protestants one more reason to vote Republican—to protect themselves against "immigrant hordes" who voted Democratic. A quarter of a century after the Civil War, the organization of party constituencies along lines of sectional, racial, and religious loyalty had been confirmed by the remarkable stability and relative balance of the multi-state network of local political institutions each major party came to possess.

2

A review of the allegiances of these Republican and Democratic constituencies reveals the extent and depth of their war-related commitments. By 1868 the white farmers of the North who had filled the ranks of the Grand Army of the Republic found a settled home in the army's political auxiliary, the Grand Old Party. The decision had been made simpler by the convenient fact that both party and army possessed the same commander in chief, Ulysses S. Grant. Throughout the North the politics of army pensions, orchestrated by loyal Republican leagues, contributed additional political adhesive. From New England to Minnesota, hundreds of small towns, as well as broad swaths of rural America, became virtual rotten boroughs of Republicanism. The original prewar coalition of free soilers, abolitionists, and Whigs which had carried Lincoln to the White House thus found in postwar sectionalism a common ground that proved far more serviceable than the controversial issue of Negro rights.

Northern blacks voted Republican for war-related reasons also, though they preferred to see the G.O.P. as an egalitarian idea—the party of emancipation—rather than as the political manifestation of a sectional army. But the passing of radical abolitionism proved rapid. The bankers, manufacturers, shippers, and merchants who had provided much of the direction for the G.O.P. from its inception soon wearied of their attempt to build a postwar party in the South based on black suffrage. As election victories in the 1860's and 1870's proved that the G.O.P. could rule with a basically Northern constituency, Negroes, their

morale declining and white radical abolitionists, their numbers
thinning, lost the intra-party debate over Southern policy. For
most white Republicans, the choice was not hard; party profes-
sionals, more enamored of election results than theories found
the politics of sectionalism—"waving the bloody shirt," in the
contemporary expression—to be far more persuasive to voters
than the elaborate defenses of black rights that were necessary
to justify Reconstruction policy in the South. As early as 1868
the Freedman's Bureau was, in effect, allowed to lapse, and the
G.O.P. thereafter gradually abandoned both the cause of the
freedman and the commitment to a "reconstructed" South that
it implied. Given the known prejudice toward blacks of a large
portion of the party's white adherents in the North, the supe-
riority of the bloody shirt as a campaign appeal was unassailable.
As Negro spokesmen grimly noted, blacks were steadily losing
their political influence—though their votes were still counted
by the reoriented Republican Party.

The orientation, it soon became apparent, belonged to business. Indeed, the decline in abolitionist zeal was more than
balanced by the triumphant spirit of business enterprise that
suffused the remodeled Northern G.O.P. Though all participants
in the world of commerce did not habitually march in perfect
political lockstep, particularly on monetary policy, a workable
hegemony within commercial ranks was fashioned in the 1870's
and 1880's as a precursor to its near total ascendancy in the
1890's. Thus, the many-faceted Republican coalition that had
come to power in 1861 became in the postwar years a much
narrower business party, closely tied to the politics of sectional
division. Only faint echoes of the multi-sectional impulses of
prewar business Whiggery remained.

If Northern blacks faced a dilemma as the party of emancipation became an engine of enthusiastic enterprise, Southern
white farmers encountered a similar problem of identification
in the restructured politics of the shattered Southland. Like
blacks, Southern whites had an emotional basis for party loyalty—
though, of course, it was to the party of the Confederacy rather
than to the party of the Union. But when federal troops marched
away at the end of Reconstruction and conservative white rule

returned, the reconstituted Democratic Party ceased to be recognizable as an institution of "the plain people." Though Southern farmers from Virginia to Texas looked upon their political home as "the party of the fathers," the postwar Democracy no longer responded to such agrarian rhythms as had existed in the times of Jefferson and Jackson. Rather, the Southern Democratic Party responded to the needs of "New South" entrepreneurs—even as the farmers who had fought in the Confederate Army continued to provide their dazed allegiance. Conceived in white supremacy and clothed with the symbolic garments of the Lost Cause, the postwar institution of business was able to attract the allegiance of white Southerners of all classes, including the small number of urban workers in the region. Indeed, in the maturing system of nostalgic "Solid South" Democratic politics, tributes to the fallen and the gallant of the Lost Cause became such ritualistic features of public speaking that almost all orations, including those at funerals, were inherently political in form—though they remained essentially nonideological in content.

These developments, of course, left Southern blacks as isolated as their counterparts in the North were. Immediately after Appomattox the war legacies that shaped the voting habits of the ex-slaves had the distinction of merging emotional loyalties with visible self-interest. The black Republican Party of the South might have been a product of the war, but it was also a logical expression of the political presence and needs of Negro people. However, by the late 1880's, with the steady deterioration of Northern Republican commitment to the civil and economic rights of freedmen, the clear political purpose underlying black allegiance to the party of Lincoln made no more sense in terms of self-interest than did the other residual war loyalties operating in the land.

3

Everywhere—North and South, among Republicans and Democrats—business and financial entrepreneurs had achieved effective control of a restructured American party system. To innovative monetary theorists, the fact was central: sectional prejudices in the 1880's and 1890's persisted as an enormous political barrier to anyone bent on creating a multi-sectional party of reform. Indeed, the mature relationship between sectionalism, issueless politics, and the business direction of both major parties not only became the animating political cause of the emergence of Populism, but the almost wholly nonideological climate created by sectional politics was also to prove the third party's principal obstacle.

By the time Benjamin Harrison moved into the White House in 1888, the postwar restructuring of the American party system had seemingly become quite settled. Though each party remained dominant in its own section and possessed lesser or greater pockets of strength in the other's bailiwick—largely for reasons of sectional, racial, or religious loyalty—both responded primarily to the needs of businessmen. Of course, neither always found it convenient to stress the matter with relentless precision.

Not political ideas but war-related emotions that had intrinsic political meaning became the central element of post-Reconstruction politics in America. While practical politicians might employ ritualistic references to the high or low tariff to dress up their party's principles, they also maintained an inventory of oratorial fire to rekindle the embers of the sectional and racial patriotism that had flamed during the war. The politics of sectionalism contained a reserve of partisan firepower that could be used against any candidate or any party attempting, through an innovative appeal to "issues," to rearrange the nation's basic postwar alignment.

Prior to the outbreak of Populism, the basic constituencies of both major parties thus remained substantially unaffected by the social and economic changes occurring in industrial America. The pervasiveness of this reality testified to the marginal impact of political theorizing.

4

It was in such a non-ideological milieu that the nation first encountered the political issue that was to transform public dialogue in the 1890's—the "financial question." The importance of the issue can scarcely be exaggerated. How money was created, and on what basis it circulated, defined in critical ways the relationships of farmers, urban workers, and commercial participants in the emerging industrial state. The answers would go a long way toward determining who controlled the rules of credit and commerce, who shared in the fruits of increasing American production, and, ultimately, how many Americans obtained that minimum of income necessary to ensure that they lived lives of some dignity. With the stakes so high, all questions about the currency were clearly not of equal weight. One of the weightiest concerned the origin of money. Whether the government issued money, or whether private bankers did, shaped the precise forms of finance capitalism to a telling degree. The resolution of this issue might well determine which occupational groups had a measure of influence over their own economic future and which did not. How much money circulated also was important—in terms of price and income levels, interest rates, and the relations of creditor and debtor classes. Central to the whole issue, of course, was a clear definition of money itself. Was it gold? Gold exclusively? Silver and gold? Did currency include paper money?

These matters tended to be discussed by monetary conservatives in rigid, moral terms. Their answers were short and clear, if often inane. It was an article of faith for them that a proper monetary system had to be based on the gold standard. Yet simple faith was not enough to win the day for hard money; the debate never seemed to go away. The currency question erupted in each decade and often reached levels that soared beyond simple morality into new and surprising realms of social, economic, and political philosophy. The stakes were high because ultimate answers, if ever obtained, would define the basic economic ground rules for American society. But if the "financial question" seemed somehow organic, it stumbled upon the stage of national politics. The origin of the issue, however, was appropriate enough: money arguments stemmed from the war, too. In fact, the controversy over "greenbacks" fitted easily within Walt Whitman's perception of the Civil War as a guide to the American future.

5

In the first year of the war the government had contracted with commercial banks for a number of war loans that had the effect of substantially increasing the currency and thus the pressure on gold reserves. Long before McClellan's army embarked for the first great Richmond campaign, the nation had quietly left the gold standard. In technical language that millions of Americans would try to comprehend over the next two generations, "specie payments" had been "suspended." Two months after the Treasury ceased paying coin for its obligations, Congress, under relentless wartime spending pressure, authorized the issuance of "legal tender treasury notes" to cover obligations. Because of the color of their ink, the notes soon became known as "greenbacks." By the end of the war some $450 million of these treasury notes were in circulation, having contributed to wartime inflation, greater commercial liquidity, and prosperity.

In orthodox financial circles favoring "gold monometallism" the postwar problem was one of ending "suspension" and achieving "resumption" by retiring the greenbacks and returning to a redeemable currency of hard money. The currency "contraction" that necessarily would follow might be painful for various members of the society, especially debtors, but only as the painful cleaning of a wound was essential to ultimate health. At the heart of the banker's approach was an understanding of gold and silver money not as a medium of exchange, but as a commodity that had "intrinsic value." In the language of orthodox "goldbugs," money "was only as good as the gold which is in it." Gold was orderly and civilized; money not backed by hard metal was "fiat money," which failed the measure of intrinsic value. It was money only because legislators, by arbitrary fiat, said it was. Such currency was essentially corrupt, and its continued use constituted a morally corrupt method of running a society.

However, bankers and other creditor-bondholders had a more specific motive for specie resumption. The currency had depreciated steadily during the war, and, having purchased government bonds then, they, understandably, looked forward to the

windfall profits to be made from redeeming their holdings in
gold valued at the prewar level. A governmental decision to
begin paying coin for its obligations would mean that, though
the Civil War had been fought with fifty-cent dollars, the cost
would be paid in one-hundred-cent dollars. The nation's tax-
payers would pay the difference to the banking community
holding the bonds. Bankers marshalled a number of moral
imperatives to support their case. They argued that they had
supported the war effort—albeit with depreciated money—by
buying government securities on the assumption that the postwar
dollar would be returned to "par." For the government to take
any action to render this assumption invalid would be unethical.
Bondholders and the Eastern financial community—the two
terms were more or less interchangeable—further argued that
resumption would encourage saving, investment, and economic
growth by assuring holders of capital that the dollar would have
"long-term stability." The country would be placed on a "sound"
footing. Finally, the banker's case was patriotic: the nation's
honor was at stake.

Some practical difficulties intruded, however. A return to hard
money could only be accomplished in one of two ways—both
quite harmful to a great number of Americans. The first was to
raise taxes and then employ the proceeds to redeem wartime
bonds and to retire greenbacks from circulation. This, of course,
would contract the currency abruptly, driving prices down, but
also depressing business severely and increasing unemployment,
perhaps to socially dangerous levels. Such a contraction was not
immediately attainable in any case, because United States prices
were so high compared to world prices that it would have been
quite profitable for Americans to redeem dollars in gold and
then buy products overseas. Any immediate attempt to "resume
specie payments" would have quickly exhausted the nation's
gold supply through an unfavorable balance of trade.

The second method of contracting the currency spread the
resulting economic pain over a longer period of time. The
government could merely hold the supply of money at existing
levels while the population and the economy of the nation
expanded, thus forcing general price levels down to a point

where it was no longer profitable to redeem paper dollars in gold to finance imports. In due course, this is what happened.

To the nation's farmers, contraction was a mass tragedy which eventually led to the Populist revolt. Although the economic relationships sound quite complex, they can be spelled out in fairly simple terms through an arbitrary numerical example. Letting ten farmers symbolize the entire population, and ten dollars the entire money supply, and ten bushels of wheat the entire production of the economy, it is at once evident that a bushel of wheat would sell for one dollar. Should the population, production, and money supply increase to twenty over a period of, say, two generations, the farmers' return would still be one dollar per bushel. But should population and production double to twenty while the money supply was held at ten—currency contraction—the price of wheat would drop to fifty cents. The farmers of the nation would get no more for twenty bushels of wheat than they had previously received for ten. Moreover, money being more scarce, interest rates would have risen considerably. A person who borrowed $1000 to buy a farm in 1868 would not only have to grow twice as much wheat in 1888 to earn the same mortgage payment he made earlier, he would be repaying his loan in dollars that had twice as much purchasing power as the depreciated currency he had originally borrowed. Thus, while contraction was a blessing to banker-creditors, it placed a cruel and exploitive burden on the nation's producer-debtors.

The debtor philosophy offered another way of stabilizing prices. By reducing the content of the dollar to one-half its prewar figure, the nation could have simply accepted the fact that the currency had lost one-half of its purchasing power, frankly and rather painlessly acknowledging that currency devaluation had taken place during the war. Granted that such a solution would remove the windfall profits that bondholders anticipated from the return of the gold standard, it also avoided the multiple hazards to the rest of the society implicit in the objectives of "sound money" bankers.

To greenbackers, the case for a fiat currency was completely persuasive because the nation needed an expanding monetary system to keep up with population growth and commercial expansion. Greenbacks were "the people's currency, elastic, cheap and inexportable, based on the entire wealth of the country." As this study of American Populism reveals, the greenback cause was a many-faceted phenomenon, sometimes put forward in arguments which were opportunist and ephemeral, but more frequently presented in a coherent analysis that attained a level of advanced social criticism.

Whatever the short-run economic equities, the greenback critique of American finance capitalism—should it ever gain a mass popular following—constituted a political issue of the first magnitude. In fact, at one time or another in the decade following the war, portions of every sector of the American population felt defrauded by bankers. Soft-money theory is most easily grasped as a political ideology grounded in a desire of non-bankers to cope with changing commercial power relationships within an industrializing society. As such, the proper place to begin is with the developing ideology itself and then pass to the movement's immediate political phases—the National Labor Union (1871), the Greenback Party (1876–84), and the Union Labor Party (1888). It is pertinent to do so because the final and most powerful assertion of the greenback critique of the American monetary system came through the mobilization of the multisectional political coalition known as the People's Party.

6

The postwar doctrines of greenbackers went back to the 1840's, to the writings of Edward Kellogg, a self-made merchant who devoted himself to a study of international finance after being ruined by the Panic of 1837. To Kellogg, currency was solely a creature of law. Far from viewing money as a commodity of "intrinsic value," Kellogg and his disciples held that "pebbles or any other material would answer the same purpose as gold and silver, if law could make them a tender for debt, and control the quantity." Since government-issue paper promised to be more manageable than stones, he advocated a fiat currency issued by a central bank. Kellogg's first full-scale work, published in 1849, outlined in its title most of the dimensions of future nineteenth-century monetary struggles: *Labor and Other Capital: The Rights Of Each Secured and the Wrongs of Both Eradicated. Or, an exposition of the cause why few are wealthy and many poor, and the delineation of a system, which, without infringing the rights of property, will give to labor its just reward.*

The impact of Kellogg's book multiplied as new editions were published in the second half of the century. Kellogg's ideas, elaborated by Alexander Campbell following the war, became the conceptual basis of the labor-greenback cause in the late 1860's. Pennsylvania economist Henry Carey, also building upon Kellogg, developed a romantic and ill-conceived philosophy of "producerism," which was designed to deliver the soft-money message with equal force both to entrepreneurial capitalists and to labor. Reduced to its political essentials, Carey attempted to oppose usurious moneylenders and foresaw a "harmony of interest" between labor and productive entrepreneurs against a common and parasitic foe, the finance capitalist.

These business greenbackers developed a measurable following among western Pennsylvania iron and steel men and others who felt gouged by Eastern commercial bankers. Yet hard-money ideas eventually prevailed in business circles even among men who had little to gain and much to lose by the ascendancy of the gold dollar. Indeed, the transformation of soft-money ironmasters into submissive, financially orthodox Republicans casts an interesting light on the social pressure for political conformity operating within affluent Northern society. The failure of the greenback cause among urban workers was more readily predictable, if only because of the fragile condition of the existing labor movement. The scarcity of labor newspapers or other organizational methods of communication within the ranks of labor deprived labor greenbackers of the means of presenting their case. Either because urban workers accepted the moral criticism of greenbacks, or because the entire matter seemed too difficult to grasp, or because they simply never heard of it, or because they developed other priorities, the soft-money creed proved easy for laboring people to resist.

Beyond its ideological difficulties, the political failure of the greenback cause is even easier to explain. The doctrine of a fiat currency fell victim, simply and forcefully, to sectionalism. The soft-money creed had nothing to offer in the way of sectional appeal; it offered instead an idea about economic relationships and then tried to prevail by explaining the idea to major party constituencies that still had emotional and cultural needs related to war-rooted loyalties. Against the sectional politics of the bloody shirt and the Lost Cause, the politics of a fiat currency recruited comparatively few regiments. Ideologically innovative, culturally embattled, and irrelevant in terms of sectional loyalties, the soft-money creed endured a fragile political life within a national party system geared to non-economic memories.

Rather effortlessly, then, Congress was able to defend the nation's honor, resuming the payments of gold for wartime bonds and gradually contracting the currency by consciously declining to increase volume in step with population growth and commercial expansion. After a measure of government hesitancy traceable to agitation by business and labor greenbackers, the amount of currency in circulation was held at a stable level through the decade of the 1870's while expanded population and production reduced price levels and spread severe economic hardship throughout the nation's agricultural districts. Hard-hit farmers were brusquely told they were guilty of "overproduction." The economic depression was both protracted and severe. Business was badly hurt and unemployment rose, but by the end of the decade the goal was reached: the United States went back on the gold standard on January 2, 1879.

7

As if these developments were not enough, still a final irony awaited the nation's currency reformers—and it was one that would cast a shadow forward, all the way into the Populist period. The greenback cause was not only defeated politically by sectionalism in the popular mind, and by "intrinsic value" homilies of sound money moralists, but even the intellectual arguments in behalf of a fiat currency became obscured by another "financial question" that intruded into the nation's politics in the 1870's. That development, which centered around a bizarre topic popularly known as "the Crime of '73," almost destroyed what little public understanding had materialized on American banking practices. To the undying distress of green-backers, the "Crime of '73" focused popular attention on an extraneous issue—the fancied need for a "remonetization of silver." The silver episode merits brief attention because, as refurbished in the 1890's by silver mineowners, it provided the rationale for the shadow movement of Populism known as "silver-fusion."

In the early 1870's, a currency bill had been introduced that, in one of its less debated features, quietly dropped the silver dollar from the nation's coinage. By the time the bill had proceeded toward enactment late in 1872, new mining methods had vastly increased the production of silver, depressing prices at a rate that indicated to all those who were aware of the meaning of such developments that silver would soon fall below par with gold. Now specie resumption threatened to become a hollow victory for the cause of sound money because the more valuable gold dollars would soon disappear from circulation and silver would necessarily be employed to redeem the wartime bonds: depreciated silver would remove some of the anticipated windfall for wartime bondholders. To knowledgeable advocates of the gold dollar, the new coinage bill to drive silver out of circulation became not merely attractive, but absolutely neces-sary. Actually, however, the importance of the silver provisions in the multi-purpose bill were not widely grasped, even in Congress, and certainly not in the country at large. Partly as a result of disingenuous explanations by its congressional sponsors, the bill attained final passage in January 1873 without even a roll call vote in the Senate. Though specie payments were not resumed at once, silver was "demonetized" and the country placed on a gold standard. Thus occurred what a generation of Americans would come to know as "The Crime of '73."

It took some time for the crime to be discovered. A gradually deepening economic depression provided the necessary aware-ness. The unexpected collapse of the famed investment banking house of Jay Cooke in the autumn of the year ignited panic in an already nervous Eastern financial community. Unemployment rose, demand softened, and wage cuts followed. As the depression worsened, prices fell, though neither as swiftly nor as far as wages did. Commodity income for farmers slumped badly. What all this meant was beyond the ken of orthodox financial thinkers of the day, but the facts seemed to be relatively simple: silver suddenly was not worth much and the country was gripped by a depression. It was at this point that the alarms sounded on the "Crime of '73." Once its congressional origins were understood, the perpetrators of the coinage bill were condemned by the public in creative legends that exposed any number of conspiracies. Counter legends, hurriedly promulgated by goldbugs, soon earned acceptance among creditors. But both views came well after the fact: the "Crime of '73" went undiscovered for almost three years before the depression had deepened sufficiently to stir inquiries and reappraisals across the land. The outcry for silver that materialized emanated from debtors, farmers, laborers, and others most vulnerable to the hardships of the depression. Yet the groundswell for silver derived less from broad public understanding of currency than from moral outrage at the apparently surreptitious means by which the bill had achieved congressional approval.

To greenback apostles, the entire controversy bordered on madness. The cry for silver was an utter delusion. To fiat-money men it was elementary that silver, like gold, was but another prop in a hard-money currency. As such, it contributed to a relatively rigid money supply that hurt the economy and, among other maladies, fostered high interest rates that benefited only moneylenders. While a measure of inflation could be achieved by coining silver on the proper terms, such terms were not being debated, were not widely understood, and were not likely to be as long as bankers continued to convince Americans that money had to have "intrinsic value."

Such greenback arguments, baffling and therefore suspect to many, drove the curious to look for corroboration in the nation's press, universities, and pulpits. With a few exceptions, the representatives of all three were rigidly orthodox on the money issue, their attitudes grounded in a deferential acceptance of the bankers' argument that currency had to have a metallic base. Though greenbackers themselves looked upon "silver Republicans" and "silver Democrats" with equal disdain, the soft-money cause had to swim upstream against the cultural presumptions of the era. As soft-money advocates looked on helplessly, the nation returned to specie payments on terms that brought deep satisfaction to goldbugs and a modest measure of "sound money inflation" to silverites.

8

Given such a remarkable range of intellectual, political, and cultural constraints, it is understandable that the political institutions created by greenbackers in the 1870's and 1880's fared rather badly. Soft-money doctrines enjoyed a brief period of prominence as a centerpiece of the politics of the National Labor Union in the early 1870's before they acquired something of the dimension of a national political presence when the Greenback Party was formed in 1876. From that point to its effective demise in 1884, the Greenback Party drew little more than ridicule from respectable society. Luhman Weller, Iowa's greenback Congressman, became widely known as "Calamity" Weller, and Texas' George Washington Jones was reduced to "Wash" Jones. "Sound money" seemed a fixture in American culture. By 1886 virtually the only legacy of a generation of soft-money agitation and party-building was a network of agrarian radicals scattered along the Western frontier from Dakota Territory to Texas. The older generation among these radicals, with memories that dated back to the first debates on the wartime greenbacks in the 1860's, were disillusioned. A goodly portion of the incoming mail to ex-Congressman Weller reflected the desperation of men who had spent years trying to employ economic arguments to overcome the sectional loyalties of the nation's voters. In overwhelmingly Republican Kansas and Iowa the farmers preferred the bloody shirt. In overwhelmingly Democratic Arkansas and Texas they preferred the "party of the fathers." When a final, desperate effort to reconstitute a third party structure collapsed in the abortive bid of the Union Labor Party in 1888, a number of greenbackers simply gave up. They were not of a mood to blame themselves. "The people are near-sighted," an embittered reformer wrote Weller. "The collapse of the Greenback Party and the poor showing made by the Union Labor Party has in my opinion destroyed all hope for a new party during the next decade, if not during a generation." That cryptic judgment (and a singularly misplaced one, considering what was about to happen) merely reinforced the deepening gloom of currency reformers. The money question might be central to the ongoing problem of economic inequity and corporate concentration in America, but the nation's political culture, reduced to a primitive level by the emotions of sectionalism, seemed thoroughly in harmony with the needs of "sound-money" bankers. "Calamity" Weller's mailbox filled with the complaints of earnest but battered reformers.

But though reformers tended to blame the "apathy" and "near-sightedness" of the people, their problem essentially was one of simple organization: they needed to find an effective instrument of recruitment. Stump speeches based on complicated greenback monetary arguments were simply not enough to affect the way people acted politically.

Help was coming, however. Even as oldtime greenbackers wrung their hands in despair, an organization new to rural America, slowly, painfully, began to shape a powerful method of mass recruitment. Indeed, in the very years of greenback decline, the new democratic vision inspired by the young organization began to affect millions of Americans. "Calamity" Weller's mailbox was merely not the place to look for it.

2

The Alliance Develops
a Movement Culture

"the morning sunlight of labor's freedom . . . ."

The agrarian revolt first stirred on the Southern frontier, then swept eastward across Texas and the other states of the Old Confederacy and thence to the Western Plains. The gathering of democratic momentum required almost fifteen years—seven within a tier of counties along the Texas farming frontier, three more to cover the rest of the state, and another five to envelop the South and West. Yet the best way to view this process of mass radicalization of farmers is not by focusing on the farming frontier from which the doctrines of insurgency emanated, but rather by discovering the humiliating conditions of life which penetrated into every farm and hamlet of the South. These conditions really illuminated the potential support for the agrarian revolt because they caused thousands to flee to the frontier and armed those who remained with a fervent desire to join the movement when it eventually swept through their region. Further, these conditions were so pervasive in their impact, shaping in demeaning detail the daily options of millions of Southerners, that they constituted a system that ordered life itself.

2

The "system" was the crop lien system. It defined with brutalizing finality not only the day-to-day existence of most Southerners who worked the land, but also the narrowed possibilities of their entire lives. Both the literal meaning and the ultimate dimension of the crop lien were visible in simple scenes occurring daily, year after year, decade after decade, in every village of every Southern state. Acted out at a thousand merchant counters in the South after the Civil War, these scenes were so ubiquitous that to describe one is to convey a sense of them all. The farmer, his eyes downcast, and his hat sometimes literally in his hand, approached the merchant with a list of his needs. The man behind the counter consulted a ledger, and after a mumbled exchange, moved to his shelves to select the goods that would satisfy at least a part of his customer's wants. Rarely did the farmer receive the range of items or even the quantity of one item he had requested. No money changed hands; the merchant merely made brief notations in his ledger. Two weeks or a month later, the farmer would return, the consultation would recur, the mumbled exchange and the careful selection of goods would ensue, and new additions would be noted in the ledger. From early spring to late fall the ritual would be enacted until, at "settlin'-up" time, the farmer and the merchant would meet at the local cotton gin, where the fruits of a year's toil would be ginned, bagged, tied, weighed, and sold. At that moment, the farmer would learn what his cotton had brought. The merchant, who had possessed title to the crop even before the farmer had planted it, then consulted his ledger for a final time. The accumulated debt for the year, he informed the farmer, exceeded the income received from the cotton crop. The farmer had failed in his effort to "pay out"—he still owed the merchant a remaining balance for the supplies "furnished" on credit during the year. The "furnishing merchant" would then announce his intention to carry the farmer through the winter on a new account, the latter merely having to sign a note mortgaging to the merchant the next year's crop. The lien signed, the farmer, empty-handed, climbed in his wagon and drove home, knowing that for the second or fifth or fifteenth year he had not paid out.

Such was the crop lien system. It constituted a new and debasing method of economic organization that took its specific form from the devastation of the Civil War and from the collapse of the economic structure of Southern society which had resulted from the war. In the aftermath of Appomattox, the people of

the South had very little capital or the institutions dealing in it—
banks. Emancipation had erased the slave system's massive
investment in human capital, and surrender had not only
invalidated all Confederacy currency, it had also engendered a
wave of Southern bank failures. Massachusetts alone had five
times as much national bank circulation as the entire South, while
Bridgeport, Connecticut, had more than the states of Texas,
Alabama, and North and South Carolina combined. The per
capita figure for Rhode Island was $77.16; it was 13 cents for
Arkansas. One hundred and twenty-three counties in the state
of Georgia had no banking facilities of any kind. The South had
become, in the words of one historian, a "giant pawn shop."

The furnishing merchants, able to get most of their goods on
consignment from competing Northern mercantile houses,
bought supplies and "furnished" them on credit to farmers,
taking a lien on the farmer's crop for security. Farmers learned
that the interest they were paying on everything they consumed
limited their lives in a new and terrible way; the rates imposed
were frequently well in excess of 100 per cent annually, sometimes over 200 per cent. The system had subtle ramifications
which made this mountain of interest possible. At the heart of
the process was a simple two-price system for all items—one
price for cash customers and a second and higher price for
credit customers. Interest of 25 to 50 per cent would then be
charged on this inflated base. An item carrying a "cash price"
of 10 cents would be sold on credit for 14 cents and at the end
of the year would bring the merchant, after the addition of, say,
33 per cent interest, a total of 19 cents—almost double the
standard purchasing price. Once a farmer had signed his first
crop lien he was in bondage to his merchant as long as he failed
to pay out, because "no competitor would sell the farmer so
much as a side of fat back, except for cash, since the only
acceptable security, his crop, had been forfeited." The farmer
rarely was even aware of the disparity between cash and credit
prices, for he usually had no basis for comparison; "many of the
merchants did a credit business so exclusively they set no cash
prices." The farmer soon learned that the prudent judgment—
or whim—of his furnishing merchant was the towering reality

of his life. Did his wife want some calico for her single "Sunday dress," or did his family need a slab of bacon? Whether he got them or not depended on the invisible scales on which the merchant across the counter weighed the central question— would the farmer's crop yield enough money to pay off the accumulating furnishing debt?

In ways people outside the South had difficulty perceiving, the crop lien system became for millions of Southerners, white and black, little more than slavery. "When one of these mortgages has been recorded against the Southern farmer," wrote a contemporary, "he had usually passed into a state of helpless peonage. . . . From this time until he has paid the last dollar of his indebtedness, he is subject to the constant oversight and direction of the merchant." The man with the ledger became the farmer's sole significant contact with the outside world. Across the South he was known as "the furnishing man" or "the advancing man." To black farmers he became "the Man."

The account books of Southern furnishing merchants present grim evidence not only of the gradations of privation between blacks and whites, but the near universality of privation. In South Carolina low farm prices forced a middle-class white farmer, S. R. Simonton, to open a credit account with the furnishing house of T. G. Patrick. While Simonton's first year's expenditures were $916.63, declining prices helped reduce his after-sale "credits" to only $307.31, leaving an unpaid balance of over $600, which he settled by note. The subsequent annual credit extended to him by the furnishing merchant did not exceed $400 per year, showing that he had suffered a drop of well over 100 per cent in his standard of living. Still, he was never able to "pay out." After seven years the debt was settled by a transfer of land to the furnishing merchant. Simonton had become a landless tenant.

Detailed records of the account of a Mississippi Negro farmer over seventeen years reveal an even grimmer degradation. Matt Brown purchased his supplies from the Jones Store at Black Hawk, Mississippi, from 1884 to 1901. Brown was not free of debt at any time in those years. He began the year 1892 with an indebtedness of $226.84 held over from previous years. At final

settlement on January 3, 1893, his obligation had increased to
$452.41. Though his credits during the year from selling cotton,
cutting wood, clearing land, and hauling for the store amounted
to $171, his expenditures for the year were $353 for household
and farm supplies. By 1895 his credit standing had diminished
to the point that his twelve-month expenditure for food totaled
$8.42. In that year he spent $27.25 on clothing, $38.30 on farm
and household supplies, 95 cents on drugs, $2.35 for a cash
advance, and $12.08 on miscellaneous supplies. Matt Brown's
account, it appears, was ultimately settled by a mortgage. A 1905
entry is for a coffin and burial supplies. The record was
permanently closed by "marking it off."

For millions of farmers of both races throughout the South,
those were the realities of life in the last half of the nineteenth
century. Southern metropolitan newspapers told farmers they
"bought too much and sold too little," but farmers who spent
$10.00 to $50.00 a year for food knew better. They could have
cited another statistic: the Southern cotton crop of 8.6 million
bales in 1890 brought $429.7 million to the farmers; the next
year's crop, 9.0 million bales, brought only $391.5 million—a
decrease of $38.2 million despite an increase in production.

New South editorialists said that the Southern farmer should
diversify and grow perishable food supplies as well as cotton,
but both the farmer and the furnishing merchant knew better.
The compelling fact was that an acre of corn produced even less
financial return than an acre of cotton did. Conservatives who
advised the farmer assumed he possessed a degree of autonomy
that he simply did not have. Furnishing merchants demanded
that their debtors plant the one certain cash crop, cotton. One
phrase defined the options: "No cotton, no credit." Moreover,
goldbug newspaper editors rarely focused on the reality that
the government's reliance on the gold standard meant deflation,
which translated into the long postwar fall of farm prices.
Farmers, caught between high interest rates and low commodity
prices, lost almost all hope of ever being able to pay out. Every
year more and more of them lost their land to their furnishing
merchant and became his tenants. Merchants began to consider
a "run" of fifty to one hundred tenants on their lands as normal.

They gradually acquired title to steadily increasing portions of the lands of the country. As thousands, then millions descended into the world of landless tenantry, the annual output expanded, but both the soil and those who worked it gradually became exhausted as a result of the desperate cycle of crop lien, furnish, cotton harvest, failure to pay out, and new crop lien.

3

It is not surprising that men were lured or driven west. Yet even the great migration that began in the 1870's did not alter the guiding principles of the new system. One Southern historian described the bitter logic which had made cotton a new king of poverty: "Let . . . the soil be worn out, let the people move to Texas . . . let almost anything happen provided all possible cotton is produced each year."

For simple, geographical reasons, "Going West" for most Southerners meant, in the familiar phrase of the time, "Gone to Texas." The phrase became so common that often only the initials "G. T. T." scrawled across a nailed-shut door were needed to convey the message. White and Negro farmers by the thousands drove down the plank roads and rutted trails of the rural South, westward across the Mississippi River to the Sabine and into the pine forests of East Texas. The quest for new land and a new start drove lengthening caravans of the poor—almost 100,000 every year of the 1870's—ever deeper into Texas, through and beyond the "piney woods" and on into the hill country and prairie Cross Timbers. There the men and women of the South stepped out into the world of the Great Plains. It was there that the culture of a new people's politics took form in nineteenth-century America.

In September 1877 a group of farmers gathered at the Lampasas County farm of J. R. Allen and banded together as the "Knights of Reliance." In the words of one of the founders, they were all "comparatively poor" and the farm organization was the "first enterprise of much importance undertaken." The overriding purpose, he later said, was to organize to "more speedily educate ourselves" in preparation for the day "when all the balance of labor's products become concentrated into the hands of a few, there to constitute a power that would enslave posterity." In his view, the farmers needed to organize a new institution for America, a "grand social and political palace where liberty may dwell and justice be safely domiciled." How to achieve such a useful "palace" was, of course, the problem.

The new organization soon changed its name to "The Farmers Alliance," borrowed freely from the rituals of older farm organizations, and spread to surrounding counties. In the summer of 1878, a "Grand State Farmers Alliance" was formed. The growing county and state structure was to be a rural organization of self-help, but workable models were in short supply. How to cope with the lien system? Some of the farmers decided politics was the answer and they tried to lead the order into the Greenback Party. But the Lampasas-based Alliance collapsed in 1880, sundered by the sectional loyalties of many members to "the party of the fathers."

The Alliance experience thus yielded its first lesson: immediate political insurgency was not the answer; too many of the poor had strong cultural memories that yoked them to traditional modes of political thought and behavior. Some means would have to be found to cut such ties before any kind of genuine people's politics was possible.

In the next two years, the organizational sprouts in frontier regions north of Lampasas, unhampered by internal political divisions, slowly took root as 120 alliances came into being in twelve counties. But the crop lien system, supported as it was by the entire structure of American commerce, proved an over-whelming obstacle. Though Alliancemen wanted to do some-thing to solve the underlying problems of agricultural credit, their fledgling cooperative efforts at buying and selling were treated with contempt by merchants. In 1883, the Alliance lost what momentum it had achieved—only thirty suballiances were represented at the state meeting.

Into this situation stepped the first Populist, S. O. Daws. A thirty-six-year-old Mississippian, Daws had developed an inter-esting kind of personal political self-respect. Raised in the humiliating school of the crop lien system, he did not believe the inherited economic folkways were fair, and he thought he had the right to say so. Indeed, Daws, a compelling speaker, was dedicated to instilling a similar kind of political self-respect in his fellow farmers. Late in 1883 the Alliance named Daws to a newly created position, that of "Traveling Lecturer," and endowed the new chief organizer with broad executive powers to appoint suborganizers and sublecturers for every county in the state of Texas.

It proved a decisive step. Within a month, Daws had energized fifty dormant suballiances into sending delegates to a state meeting where the entire cooperative effort was put under review. If merchants practiced monopolistic techniques by refusing to deal with the Alliance, perhaps Alliance members could reply in kind. A "trade store" system was agreed upon wherein Alliance members would contract to trade exclusively with one merchant. The range of discussion was broad: it extended to the role of Alliance county business agents and Alliance joint stock companies, to the opposition of cotton buyers and the resistance of manufacturers who refused to sell to the Alliance except through middlemen, and even to the refusal of townsmen to sell farmers land for Alliance-owned cotton yards where they might store their crops while awaiting higher prices. Determined to test the trade store system, Alliance delegates dispersed from the February meeting with new hope. Daws's efforts had been so impressive that his office and his appointment powers were confirmed by the convention.

The spring of 1884 saw a rebirth of the Alliance. Daws traveled far and wide, denouncing credit merchants, railroads, trusts, the money power, and capitalists. The work-worn men and barefooted women who gathered to hear the Alliance lecturer were not impressive advertisements for the blessings of the crop lien or the gold standard. Such audiences did not require detailed proof of the evils of the two-price credit system and the other sins of the furnishing merchants. They had known "hard times" all their lives. But Daws could climax his recitation of exploitation with a call for a specific act: join the Alliance and form trade stores. Were monopolistic trusts charging exorbitant prices for fertilizers and farm implements? Join the Alliance and form cooperative buying committees. Did buyers underweigh the cotton, overcharge for sampling, inspecting, classifying, and handling? Join the Alliance and form your own cotton yard. The new urgency was evidenced not only by a steady increase in suballiances to over 200, but also in the style of the men who materialized as presidents and lecturers of the recently organized county structures—men like Daws: articulate, indignant, and capable of speaking a language the farmers understood.

4

Foremost among them was a thirty-four-year-old farmer named William Lamb. Beyond his red hair and ruggedly handsome appearance, there was at first glance little to distinguish Lamb from scores of other men who had come west to escape the post-Civil War blight of the South. A Tennessean, Lamb had arrived on the frontier at the age of sixteen, settling first near what was to become the town of Bowie in Montague County. Like most rural Southern youths, Lamb had had almost no formal education— a total of twenty-five days. After working as a hired man, he married, rented land, and began farming in Denton County. In 1876 he "preempted" a farmstead in Montague County, living alone part of the time in a rude log hut until he could clear the land and build a homestead. He farmed by day and read by night, acquiring the strengths and weaknesses of self-taught men. When he spoke publicly, his sentences were overly formal, the syntax sometimes losing its way; in the early days, when he began to write in behalf of political causes, the strength of his ideas had to overcome impediments of grammar and spelling. To Daws, who was seeking other men of energy, those faults were no liability; the younger man had strength of mind, and Daws realized it when the two men met in 1884. Lamb knew the extent of the agrarian disaster in the South, and he had an urge to do something about it. He was ripe for the missions Daws was ready to give him. By the time of the 1884 state meeting, Lamb's work in organizing suballiances in his own county had attracted enough attention that a new statewide office was created specifically for him. He was made "state lecturer" while Daws continued in the role of traveling lecturer.

   William Lamb emerged in 1884–85 as a man of enormous energy and tenacity. As president of the Montague County

Alliance he had organized over 100 suballiances by October 1885, a record that eclipsed even Daws's performance in his own county. Lamb soon surpassed his mentor in other ways. As a spokesman for farmers, the red-haired organizer thought in the broadest tactical terms. He early saw the value of a coalition between the Alliance and the Knights of Labor, which was then beginning to organize railroad workers in North Texas. Lamb also pushed cooperative buying and selling, eventually becoming the Alliance's most aggressive advocate of this program. And, after the business community had shown its hostility to Alliance cooperatives, he was the first of the Alliance leaders to react politically—and in the most sweeping terms. Encouraged by his association with Daws and the Alliance, Lamb demonstrated that he, too, had developed a new kind of personal political self-respect.

The increasing momentum of the idea of the Alliance cooperative inexorably shaped the lives of Daws and Lamb as they, in turn, shaped the lives of other men. They spent their lives in political reform, and in so doing became allies, exhorted their comrades, crowded them, sometimes challenged them, and performed the various acts that men in the grip of a moral idea are wont to perform. They influenced each other steadily, the impetus initially coming from Daws, as he introduced Lamb into an environment where the younger man's political horizons could be broadened, then from Lamb, as he pressed his views to the point of crisis and carried Daws along with him until the older man could use his influence and creativity to resolve the crisis and preserve the forward momentum of the organization.

In the personal relationship between William Lamb and S. O. Daws a rhythm was discernible: along the course they set toward independent political action, Daws was always a step or two behind; at the moment of triumph, when the third party was formed in the South, Lamb, not Daws, held the gavel.

5

The young organizer learned in 1884–85 that cooperative buying and selling was easier to plan at country meetings than to carry out. Town merchants opposed cooperative schemes, as did manufacturers and cotton buyers. Indeed, the entire commercial world was hostile to the concept. Cooperation was not the American way; competition was.

But if the new movement did not invariably achieve immediate economic gains, the cooperative idea spurred organizing work. The 1885 state meeting of the Alliance was the largest gathering of farmers ever held in Texas to that time. The order adopted a program calling on all members "to act together as a unit in the sale of their product" and to that end moved to have each county alliance set apart a special day for selling. Thus, Alliancemen began what they called "bulking." These mass cotton sales were widely advertised and cotton buyers contacted in advance, for the Alliance sought a representative turnout of agents who might themselves engage in a modicum of competition.

The cooperative movement clearly stirred a new kind of collective self-confidence among Alliance farmers: county trade committees amassed 500, 1000, and in some instances as many as 1500 bales of cotton at Alliance warehouses. Because of the convenience of bulking, particularly to foreign buyers, the trade committees asked for premium prices, 5 to 10 cents per 100 pounds above prevailing levels. Though results were mixed, the successful sales gave the farmers a sense of accomplishment. After one mass sale at Fort Worth had brought 5 cents per 100 above what individual farmers had received, Alliancemen were ecstatic. As one metropolitan daily reported it, their "empty wagons returned homeward bearing blue flags and other evidence of rejoicing." It did not take long for such stories to spread through the farming districts, and each success brought the Alliance thousands of new members. Cooperation worked— in more ways than one.

But though bulking helped, and was a marvelous boon to the self-respect of individual farmers and to the growing collective self-confidence represented by the Alliance movement, it did not eliminate the furnishing merchant, nor did it fundamentally alter the two-price credit system. Alliancemen slowly discovered they needed to establish an internal method of communication which would arm the Alliance purchasing agent with hard facts so that he could inform manufacturers of the size and value of the Alliance market for their products. Late in 1885 the Alliance moved to centralize its cooperative procurement effort by naming its own purchasing agent. The Alliance state president appointed William Lamb as the official "Traveling Agent" to represent the order "for sale of farm implements and machinery through the state where the Alliance is organized."

Though the long-term effects of that decision were to change the whole direction of the Farmers Alliance, the results in the short run were disappointing. Despite zealous efforts, Lamb failed in his attempts to establish direct purchasing arrangements with commercial America. It was the credit problem again. "I can furnish wagons in car-load lots cheap for cash, but as yet I have no offer on time," Lamb reported to his Alliance brethren. The young lecturer pondered the implications of the dilemma and decided the ultimate answer had to lie in cooperative manufacturing efforts by the Alliance itself! His plan, based on his loss of faith in the good will of merchants, scarcely constituted a serious threat to the commercial world, but it represented the first public hint that his political perspective was shifting.

And he was not alone. The "Erath County Alliance Lumber Company," with 2800 Alliance members, announced "we stand united. . . . We can purchase anything we want through our agent, dry goods, and groceries, farm implements and machinery. We have a market for all of our wheat, oats and corn." Meanwhile, the Tarrant County Alliance announced the organization of its forty-sixth suballiance and claimed a total of 2000 members throughout the county.

"The Farmers Alliance is making its power felt in this state now," a country newspaper reported approvingly. And, indeed, by October 1885, the order reported that the total number of suballiances stood at 815. Each suballiance had its lecturer; each county had its county Alliance and county lecturer. The message of agrarian self-help now sounded from hundreds of platforms.

Yet the outside world knew almost nothing of the growing energy of the Alliance cooperative movement. Few in the power centers of Texas or the nation were remotely aware of the gathering impatience among the members of a rural organization

calling itself the Farmers Alliance. Ironically, it was the conservatives of the Patrons of Husbandry—better known as the
Grange—who became the first to sense the new mood in the
rural districts of the state. The route to this perception was a
painful one. Initially, Grangers had watched with equanimity
the early stirrings of the Farmers Alliance and, courteously, if
condescendingly, had rebuffed suggestions for cooperation. But
soon complacency within the Grange leadership gave way to
alarm, then to anger, and finally to helpless, private denuncia-
tion. Nothing availed. Within thirty-six months the Grange was
all but obliterated in Texas.

The order probably deserved a less ignominious fate. The
Grange had ideas about cooperation too, though they were very
cautious ideas. It had spread across the Midwestern plains in
the early 1870's, but when the new form of agrarian assertion
manifested itself in political action, the order suffered internal
divisions and lost its organizing momentum. The basic problem
was that the Grange system of cash-only cooperative stores,
based on the English Rochdale plan, simply failed to address
the real ills of farmers. Because of the appreciating value of the
currency—and attendant lower farm prices—most farmers simply could not participate in cash-only cooperatives. They did
not have the cash. The unpleasant truth was that a cash store
was of little help if one had no money and had to deal with
credit merchants. By failing to alleviate the farmers' distress,
the Grange soon lost the bulk of its membership. The Farmers
Alliance, in contrast, had begun to grow: its membership total
of 10,000 in the summer of 1884 had soared to 50,000 by the
end of 1885.

6

It is appropriate, at the moment of Alliance ascendency, to
explore the process that had produced an emerging mass
movement of farmers. Tactically, the rise of the Alliance was a
result of its determination to go beyond the cash stores of the
Grange and make pioneering efforts in cooperative marketing
as well as purchasing. The cooperative effort was helpful because

it recruited farmers by the thousands. But in a deeper political sense the Alliance organization was experimenting in a new kind of mass autonomy. As such, it was engaged in a cultural struggle to redefine the form and meaning of life and politics in America. Out of the individual sense of self of leaders like S. O. Daws and William Lamb the Alliance had begun to develop a collective sense of purpose symbolized by the ambitious strivings of scores of groups who were anxious to show the world why they intended to "stand united." Inexorably, the mutually supportive dynamics inherent in these individual and collective modes of behavior began to produce something new among the huge mass that Alliancemen called "the plain people." This consisted of a new way of looking at society, a way of thinking that represented a shaking off of inherited forms of deference. The achievement was not an easy one. The farmers of the Alliance had spent much of their lives in humiliating circumstances; repeated dealings with Southern merchants had inculcated insecurity in generations of farming people. They were ridiculed for their poverty, and they knew it. They were called "hayseeds" and they knew that, too. But they had also known for decades that they could do nothing about their plight because they were locked into the fabric of the lien system and crushed by the mountain of interest they had been forced to pay. But now, in their Alliance, they had found something new. That something may be described as individual self-respect and collective self-confidence, or what some would call "class-consciousness." All are useful if imprecise terms to describe a growing political sensibility, one free of deference and ridicule. Such an intuition shared by enough people is, of course, a potentially powerful force. In whatever terminology this intuition is described, it clearly represents a seminal kind of democratic instinct; and it was this instinct that emerged in the Alliance in 1884–85.

In the succeeding eighteen months their new way of looking at things flowered into a mass expression of a new political vision. We may call it (for that is what it was) the movement culture of Populism. This culture involved more than just the bulking of cotton. It extended to frequent Alliance meetings to plan the mass sales—meetings where the whole family came,

where the twilight suppers were, in the early days, laid out for
ten or twenty members of the suballiance, or for hundreds at
a county Alliance meeting, but which soon grew into vast
spectacles; long trains of wagons, emblazoned with suballiance
banners, stretching literally for miles, trekking to enormous
encampments where five, ten, and twenty thousand men and
women listened intently to the plans of their Alliance and talked
among themselves about those plans. At those encampments
speakers, with growing confidence, pioneered a new political
language to describe the "money trust," the gold standard, and
the private national banking system that underlay all of their
troubles in the lien system.

How is a democratic culture created? Apparently in such
prosaic, powerful ways. When a farm family's wagon crested a
hill en route to a Fourth of July "Alliance Day" encampment
and the occupants looked back to see thousands of other families
trailed out behind them in wagon trains, the thought that "the
Alliance is the people and the people are together" took on
transforming possibilities. Such a moment—and the Alliance
experience was to yield hundreds of them—instilled hope in
hundreds of thousands of people who had been without it. The
successes of the cooperative effort gave substance to the hope,
but it was the hope itself, the sense of autonomy it encouraged
and the sense of possibility it stimulated, that lay at the heart of
Populism. If "the Alliance was the people and the people were
together," who could not see that the people had created the
means to change the circumstances of their lives? This was the
soul of the Populist faith. The cooperative movement of the
Alliance was its source.

In 1884–85, the Alliance began developing its own rhythm of
internal "education" and its own broadening political consciousness among leaders and followers. The movement culture would
develop its own mechanism of recruitment (the large-scale credit
cooperative), its own theoretical analysis (the greenback interpretation of the American version of finance capitalism), its own
solution (the sub-treasury land and loan system), its own symbols
of politics (the Alliance "Demands" and the Omaha Platform),
and its own political institution (the People's Party). Grounded
in a common experience, nurtured by years of experimentation

and self-education, it produced a party, a platform, a specific new democratic ideology, and a pathbreaking political agenda for the American nation. But none of these things were the essence of Populism. At bottom, Populism was, quite simply, an expression of self-respect. It was not an individual trait, but a collective one, surfacing as the shared hope of millions organized by the Alliance into its cooperative crusade. This individual and collective striving generated the movement culture that was Populism.

7

The first spectacular flowering of this culture came in 1886 against the backdrop of a bitter labor controversy that has come down in history as the "Great Southwest Strike." A bold new assertiveness on the part of emerging Alliance spokesmen was essential to this flowering, but less so than the new cooperative experiences within the membership that had the effect of sanctioning radicalism. These dynamics affected conservative agrarians, an embattled labor organization, a unilaterally called and richly controversial boycott, the actions of one of the nation's most notorious robber barons, the most violent labor struggle in the history of the frontier South, the reactions of the metropolitan press, and, ultimately, the structure and purpose of the Farmers Alliance itself. Given its complexity, the emergence of the mass behavior the nation was later to know as Populism is perhaps most easily perceived as it happened—one step at a time.

For decades as the nation industrialized following the Civil War, American industrial workers, in ways not dissimilar from those of farmers, had groped for ways to defend themselves against the forms of exploitation associated with the new corporate system. The most numerically significant of these efforts developed through an institution known as the Knights of Labor. In 1885, the Knights achieved, or thought they had achieved, a critical breakthrough. They had forced Jay Gould, railroad magnate and guiding spirit of the Missouri-Pacific lines, to honor a union contract. The cry "we made Jay Gould recognize us" was compelling, and in 1885–86 the Knights used it to multiply their national membership from 100,000 to 700,000. But in the spring of 1886, Jay Gould, through his general manager, H. M. Hoxie, moved to crush the union by precipitating a conflict. The struggle began when Hoxie fired a union spokesman in Texas for missing work while attending a union meeting—after the railroad had given him permission to do so. The union's members, their very right to existence challenged, rallied to defend their organization. The strike began in East Texas and quickly spread across the West. Gould moved to keep his railroads operating through strikebreakers who had been deputized as peace officers by numerous cooperating public officials. The workers were not of a mood to accept this quietly, for the very future of their union was at stake.

From beginning to end, the Great Southwest Strike was a series of minor and major battles between armed strikers and armed deputies and militiamen, interspersed with commando-like raids on company equipment by bands of workers. Thousands were indicted, hundreds were jailed, and many died. Workers "killed" railroad engines by displacing various connections—in the words of one indictment, "in such a manner as to unfit said engine for use by said company." Major newspapers denounced the strikers, suggested that their grievances were imaginary, praised Hoxie's "magnanimity," and repeatedly predicted the imminent return of the men to their jobs. It was within those shifting emotional currents that Alliance radicalism emerged.

It turned out that William Lamb's experiences as the Alliance's first purchasing agent had brought him to a new plateau of analysis concerning the farmer's place in industrial society. As constituted, cooperatives could not work because the commercial world possessed a monopoly of the money supply and effectively withheld credit. Unless farmers did something truly bold, they were locked into the lien system forever. For Lamb, the solution was a national farmer-labor coalition to restructure American politics. The Alliance had to get into politics, but first it had to cement its relations with the workers of the Knights of Labor. The "plain people" needed to be united. To that end, Lamb, as president of the Montague County Alliance, issued a boycott proclamation in support of the Knights of Labor. The Alliance state president, Andrew Dunlap, immediately replied that Lamb had no authority to order a statewide boycott and the Alliance state secretary followed with an official denial that the Alliance had "anything to do with the boycott." A battle was on to define the meaning and purpose of the farmers movement.

Lamb's boycott was only a piece of a larger question that tested the political orientation and sophistication of the entire membership. Arrayed on one side were the Alliance's top officialdom, including the editor of its state journal, J. N. Rogers of the Jacksboro *Rural Citizen*. Rogers possessed some traditional American cultural presumptions. He wrote that advocates of boycotts were "busy bodies in other men's business. When inexperienced men take the management of commerce into their hands, then woe to the commerce and business interests of our land . . . continue the strikes and we will continue the hard times." To Rogers, as to President Dunlap, it was unthinkable that the farmers of the Alliance would place their promising organization in jeopardy by allying it with the controversial Knights of Labor, an organization that was daily being pilloried in the metropolitan press. But, as often happens in mass institutions, titular leaders can gradually drift a good distance from the sentiments of their members. The cooperative crusade had taught many farmers to think in a new way about credit merchants, bankers—and railroad impresarios like Jay Gould. And William Lamb, as state lecturer and principal purchasing officer of the Alliance, was in touch with those sentiments in the suballiances much more intimately than either the state president or the editor of its official journal. Lamb accepted battle.

In an open letter to the Alliance membership, Lamb set out to clarify Rogers's thinking on the issues at stake. In the process, he directly challenged Alliance President Andrew Dunlap and brought all the issues confronting the Alliance into as sharp a focus as his skill as a writer permitted him. These issues included the failure of the cooperative buying program, the resulting need for political action, the advisability of political coalition with the Knights of Labor, and the relationship of the Alliance rank and file to its leadership. His letter was nothing less than a review of the purpose of the farmers movement.

He began with a blunt appraisal of the politics of the *Citizen:*
"We did not expect to please all members of the Alliance and
especially the Ed. of our official journal, as we have never yet
seen where he has come out and shown his true colors editorially
on the labor question." And he redefined the relationship
between members of the order and the state president:

> we think all members should show the world which side they
> are on . . . and if our State President don't wish to do so, we
> know that it will not kill him to say so, and do it in a kind
> manner, and not tell us that we are unwise or fools. . . . We
> also know that it is the duty of the President to preside with
> impartiality. . . .

Then Lamb analyzed the immediate policy question. His
experience as the order's central purchasing agent shaped his
attitude. "The writer of this article has many letters from
factories today, to the effect that they will not trade with the
Farmers Alliance except through their [own] agents." He was
anxious for Alliancemen to ponder the implications of this
reality and to set a new course in response. In one long,
breathless sentence he indicated that he had made his own
decision.

> Knowing that the day is not far distant when the Farmers
> Alliance will have to use Boycott on manufacturers in order
> to get goods direct, we think it is a good time to help the
> Knights of Labor in order to secure their help in the near
> future, knowing as we do that the Farmers Alliance can't get
> a plow today except it come through two or three agents, and
> feeling assured that the only way we can break this chain is to
> let them alone, and let them make their plows and wear them
> out themselves.

Having explained the situation as he saw it, Lamb addressed
what he regarded as the leadership's inadequate response.

> I feel satisfied that I know more about what is going on against
> us than the State President of the Alliance or our Ed., either.
> . . . Would say that one Sub-Alliance in our county killed a
> weekly paper sometime ago, and we are looking forward for
> men that will advocate our interests, those who are working
> against us are no good for us.

Lamb's cutting criticism was grounded in moral outrage, and in closing he made his underlying attitude explicit:

> we know of a certainty that manufacturers have organized against us, and that is to say if we don't do as they say, we can't get their goods. . . . Then for it to be said that we are unwise to let them alone, we can't hold our pens still until we have exposed the matter and let it be known what it is we are working for.
>
> [Signed] W. R. Lamb.

Alliance radicalism—Populism—began with this letter. The phrases, directed at so many different yet related targets, formed a manifesto of insurgent thought, summarized in the conclusion that "those who are working against us are no good for us." The obstacles that had to be overcome were clearly, if not always grammatically, delineated: editors who failed to show their "true colors on the labor question," presidents who did not know "what is going on against us," and people who failed to realize that it was "a good time to help the Knights of Labor in order to secure their help in the near future."

But Lamb's open letter to the Alliance journal was more than a political statement. His argument reflected a new conception about the farmers' place in American society. The farmer as producer-entrepreneur and small capitalist—the "hardy yeoman" of a thousand pastoral descriptions—is nowhere visible in Lamb's view. This traditional portrait, dating from a simpler Jeffersonian era and still lingering in the social tradition of the Grange, was patently out of place to a man who saw society dominated by manufacturers and their "agents." To Lamb, the farmer of the new industrial age was a worker, and the "labor question" was the central issue on which editors, presidents, and others were expected to show their "true colors." It was a simple matter of self-respect. Once such a perspective was attained, it was axiomatic that the organized farmers of the Alliance should join forces with the organized workers of the Knights of Labor. In the new era of business centralization, farmers who continued to aspire to friendship and parity with the commercial world were simply failing to comprehend "what is going on against us." Alliancemen had to put aside such naïveté: "all members

should show the world which side they are on." Lamb was asking
the farmers of Texas to achieve a new evaluation of what they
were so that they might manage a fresh assessment of their
relationship with other participants in industrial American society. Did the Alliance have the collective self-confidence to stand
against the established order? Lamb's sense of moral authority—
"we can't hold our pens still"—ensured that he would carry the
struggle over Alliance tactics to a conclusion.

Once joined, the battle was fought with considerable vigor.
Alliancemen defined themselves politically, were defined by
others, and began to ponder the differences between the two.
In the process, what was still a regional organization of farmers
developed a new sense of mission. Various alliances passed
resolutions favoring the boycott, while Rogers asserted in the
Citizen that it "was striking at the fundamental principles of
American liberty. It is putting burdens on the farmers that they
are not able to bear. It is tyranny." Late in March Alliance
President Dunlap assembled part of his correspondence with
Lamb into one long account and dispatched his own summary
to the *Citizen* and to other papers in North Texas. He "individually and officially" declared the boycott order by "this man,
W. R. Lamb" to be null and void. He called Lamb's unauthorized
proclamation of the boycott an act of "unblushing impudence."

But support for the workers was growing in dozens of Alliances
and went beyond boycotting to include joint political meetings
with Knights of Labor assemblies and direct aid to strikers.
Alliances in an East Texas county contributed farm produce
and even money to strikers, and the practice soon spread to
other Alliances along the trackage of the railroad. While outside
observers pondered the implications of organized farmers bringing food to organized strikers, Alliancemen and Knights joined
in a "Laboring Men's Convention" in President Dunlap's home
county, where they laid the groundwork for an independent
political ticket.

The state's press drew various conclusions from these evi-
dences of Alliance solidarity with the strikers. The Waco *Examiner*
blamed the Alliance for the strike's continuation: "But for the
aid strikers are receiving from farmers alliances in the state and

contributions outside, the Knights would have gone back to work long ago." The Dallas *Mercury*, strongly pro-labor, was pleased: "The Farmers Alliances of Texas are generally regarded as the spinal column of this great railroad war." The Austin *Statesman* worried about the political implications. "Unless some eruption occurs between the Knights of Labor and the Farmers Alliance," reasoned the paper, "the affairs of Texas stand a good chance of falling into the hands of those organizations at the state elections next fall."

But Alliance lecturers were not heeding city journalists. Said one: "Thanks to Providence for sending us so many good lawyers, honest bankers, and unsophisticated editors who are so good and kind to bestow their efforts to guide and direct us in the ways of prosperity." He added in anger, "Show us wherein either of the great political parties reflect any of the features of the honest face of toil . . . Shame on your boasted institutions of liberty." A new way of looking at things, clearly.

But the Knights, despite all they could do, found no way to stop Jay Gould from hiring strikebreakers. It was a problem that was to defeat the American labor movement for another half-century. As railroad traffic, protected by armed guards, returned to normal and defeat loomed before the Knights of Labor, a North Texas Alliance attacked Rogers for trying "to arouse the Farmers Alliance against the Knights of Labor." In a resolution that expressed the mid-strike mood of the order's emerging left wing, the farmers concluded: "We know if the Knights of Labor could receive all they deserve [of] the support of all the laboring classes, they would in the near future bring down the great monopolists and capitalists and emancipate the toilers of the earth from the heavy burdens which they now have to bear on account of organized capital."

8

As tempers worsened, Alliancemen could agree on one thing— the order was growing. In February, state officials placed the membership at "about 55,000," and in March, as controversy over the boycott heightened, the *Citizen* reported that Alliances were "on the increase very fast in the whole state." But the growing membership of the Alliance was balanced by the losses in the ranks of the Knights of Labor. In May, the Great Southwest Strike came to an end. No formal settlement was necessary. The members of the union were utterly destitute, and the union itself had been crushed. Most of the leaders and hundreds of members were in jail. Those of the survivors who were acceptable to the Gould management went back to work on company terms. The Knights of Labor never recovered from its defeat at the hands of Jay Gould and H. M. Hoxie. In four years its 700,000 national membership dwindled to 100,000.

Within the Alliance, however, the organizing of farmers accelerated. New leaders emerged, some of them possessed of an activism that should have given Editor Rogers pause. On the old farming frontier, the Alliance movement was led by Evan Jones, an articulate farmer destined for national fame in the agrarian revolt. One resolution, written by Jones and passed unanimously, fairly bristled:

> Whereas combined capital by their unjust oppression of labor are casting a gloom over our country and . . . Whereas we see the unjust encroachments that the capitalists are making upon the different departments of labor . . . we extend to the Knights of Labor our hearty sympathies in their manly struggle against monopolistic oppression and . . . we propose to stand by the Knights.

Throughout May the impulse toward political independence spread through the North Texas Alliances as the order's top leadership tried various experiments in an effort to cope with the restless energy. President Dunlap was finally reduced to issuing public pleas to the Alliance membership to abandon the boycott he had tried to halt since January. He got nowhere.

The organizational father of the movement, S. O. Daws, took an entirely different approach. At the height of the Dunlap-Lamb imbroglio, Daws attempted to serve as moderator: "Capital is thoroughly organized, but when the laboring class begins to organize, they call it communism and other hard names, which is unjust. I am proud the morning sunlight of labor's freedom is shining in the political horizon of the east." But as the "Alliance in politics" issue reached the boiling point, Daws decided the order's activists had to have greater freedom of action if the Alliance was not to tear itself to pieces. His solution was sufficiently conservative in its appearance and sufficiently radical in its specific uses that it became the tactical foundation upon which third-party advocates always thereafter rested their case. Daws presented his political redefinition in two sentences: "There is a way to take part in politics without having it in the order. Call each neighborhood together and organize anti-monopoly leagues . . . and nominate candidates for office." The economy of the message did not lessen the heavy ideological load it carried. At first glance the advice seemed not unlike the traditional "nonpartisan" position which, since Lampasas, had consisted of instructing Alliancemen to vote as they pleased as individuals but to keep the order out of politics. But the phrase "anti-monopoly leagues" imparted a distinct ideological definition to the kind of political effort Alliancemen should fashion in their "neighborhoods." Democratic conservatives would hardly flock to anti-monopoly leagues. Presumably, the farmers had learned something about the politics of Democrats in the Great Southwest Strike. If, during that struggle, the strikers had been able to discover any comradeship with the state police—who were Democrats all, of course—any testimony of such affection had been hard to hear above the gunfire. Democrats could not be the friends of railroad workers because the old party was dominated by railroad corporations.

The practical political effect of the three phrases—"call each neighborhood together . . . organize anti-monopoly leagues . . . and nominate candidates for office"—added up to a one-sentence mandate for insurgent political action. But S. O. Daws was not merely a radical activist, he was an organizational leader. In summarizing his interpretation of what the order's stance should be, Daws provided additional protection for the newly defined momentum, once again using familiar Alliance terminology.

> We believe in the farmer voting himself, and not being voted by demogogues. . . . Beware of men who are trying to get politics into this non-political organization, instead of trying to devise means by which the farmers may have the opportunity to emancipate themselves from the grasp of political tricksters.

This interpretation radically redefined the evils threatening the order. The Alliance was to avoid demagogic politicians—but not farmers who were "trying to devise means . . . to emancipate themselves." "Political tricksters" were to be feared—but farmers who voted their own minds were not. His analysis turned on the same perception of the role of farmers in an industrial society that William Lamb had articulated in his defense of his boycott proclamation. Daws was recommending to the farmers a course of political action that corresponded with the new definition of their class position and their growing self-respect. In 1886, Daw's ideological achievement was internally consistent, politically artful, and highly germane to the immediate demands of the agrarian movement. With a directness and a diplomacy no Grange leader was ever able to achieve, he had confronted the central dilemma of a "reform" organization saddled with a "nonpolitical" tradition. In so doing, he redefined politics in a way that provided necessary breathing room for the rival factions at a time when the order was beginning its greatest period of growth. The speed with which the Alliance moved to exploit this opportunity was dramatic: less than three months later the Alliance, in convention assembled, promulgated an aggressive seventeen-point political program that soon became famous in the agrarian movement as the "Cleburne Demands."

The ideological momentum toward the explosion at Cleburne came from two events: William Lamb's boycott of January, which threatened the unity of the Alliance, and S. O. Daw's political redefinition of May, which preserved it on new grounds. Yet the most significant development within the Alliance in the spring of 1886 was not that Lamb initiated certain events or that Daws responded to them, but rather that thousands of farmers began to express their hopes for themselves through tangible political acts.

9

The emergence of William Lamb's radical leadership and the development of the Daws formula for political action gave direction to the organizational consolidation by the Alliance of its farmer constituency in Texas in 1886. In the bitter aftermath of the Great Southwest Strike, in the face of growing newspaper awareness and concern, the order's growing corps of lecturers combed rural Texas and recruited farmers by the tens of thousands. After nine years of trial and error, the Alliance organizers had a story to tell. Their organizational framework had evolved since the founding days of 1877. Their message of economic cooperation drew from experiences dating back to the early 1880's. The order's relationship to politics had become increasingly focused since 1884. By 1886, Alliance lecturers were speaking in behalf of a maturing idea. As the summer began, the order counted 2000 suballiances and over 100,000 members.

As the Alliance came to envelop whole regions of the state, its leaders reacted in new ways to the new facts of organizational maturity. The top leadership—the president, state secretary, and state treasurer—acquired the coloration of centrists. They no longer engaged in harsh personal attacks on Alliance radicals. But the order's county and local leaders changed, too. Closest of all to the economic anguish at the bottom of Texas society, they became increasingly activist. Day after day, the local lecturers traveled through the poverty-stained backwaters of rural Texas and met with farmers in country churches or crossroads schoolhouses. The small stories of personal tragedy they heard at such meetings were repeated at the next gathering, where, in an atmosphere of genuine shared experience, they drew nods of understanding. The most astute organizers soon learned that farmers were more likely to link their own cause to another, larger one whose spokesmen knew and understood their griev-ances. The difficulty was the the lecturers themselves were altered by these experiences. They were, in effect, seeing too much. Hierarchical human societies organize themselves in ways that render their victims less visible; for a variety of reasons, including pride, the poor cooperate in this process. But the very duties of an Alliance lecturer exposed him to the grim realities of agricultural poverty with a directness that drove home the manifest need to "do something." Repeated often enough, the experience had an inexorable political effect: slowly, one by one—and in many instances unknown as yet to each other—local lecturers came to form a nucleus of radicalism inside the movement. As one new rebel defined matters, "we have an overproduction of poverty, barefooted women, political thieves and many liars. There is no difference between legalized robbery and highway robbery. . . . If you listen to other classes, you will have only three rights . . . to work, to starve, and to die."

In county after county, Alliance meetings begun in prayer and ritual ended in political speeches, many of them delivered by the suballiance chaplain, who frequently doubled as lecturer and organizer in spreading the new social gospel. Voices repeating the familiar doctrine of nonpartisanship became difficult to hear above the mounting din of insurgent political language. To the dismay of the order's old guard, Alliancemen by the thousands had gone "in politics" on the eve of the Alliance state convention in Cleburne. Whatever the majority of organized farmers now thought, the state would soon know, for the years of Alliance anonymity were over.

10

In August 1886 metropolitan reporters descended on the little farming community of Cleburne, near Dallas, to learn what they could of the large and strangely aggressive organization that called itself the Farmers Alliance. Newsmen quickly discovered that the order had little to offer in sartorial elegance. Alliancemen, they reported, looked "grangy." The appearance of the delegates, however, did not diminish their obvious seriousness.

The fashioning of a proposed political program became the work of the general "ways and means" committee, traditionally known in Alliance circles as "The Committee for the Good of the Order." The committee first announced that it had changed its name to the "Committee for the Good of the Order and Demands." The meaning of the additional two words was lodged in the committee's report itself. The proposals were not like the "petitions" which had historically characterized Grange resolutions touching on legislative matters. Rather, each of the committee's proposals was preceded by two words: "We Demand."

The committee presented its report to the convention, and a long debate followed. It was intermittently interrupted for other business, including the election of officers. The order's activists, perhaps aware that they had put enough before their colleagues for one convention, did not contest the official slate, with its nice balance of centrists and radicals. Andrew Dunlap was quietly reelected in the midst of the very internal debate over politics he had tried unsuccessfully to forestall.

But though the schism did not extend to personalities, the Alliance made a fundamental shift on another matter more directly related to the political question. The delegates, in executive session, voted to dispense with J. N. Rogers and to reject his *Rural Citizen* as their journal. They then placed the imprimatur of the Alliance on the most outspoken anti-monopoly newspaper in Texas, the Dallas *Mercury*.

With these matters decided, the delegates focused on the committee's seventeen "demands" addressed to the governments of Texas, the United States, and to President Dunlap himself. The farmers sought "such legislation as shall secure to our people freedom from the onerous and shameful abuses that the industrial classes are now suffering at the hands of arrogant capitalists and powerful corporations." Five of them dealt with labor issues, three with the power of railroads, and two with the financial problem. Of six demands relating to agricultural matters, five focused on land policy and the sixth on commodity dealings in agricultural futures. The final demand, directed to Dunlap, could be classified as "educational."

The committee placed first on its list a demand for the recognition of trade unions and cooperative stores. Other labor planks called for the establishment of a national bureau of labor statistics "that we may arrive at a correct knowledge of the educational, moral and financial condition of the labor masses," the passage of an improved mechanics lien law "to compel corporations to pay their employees according to contract, in lawful money," and the abolition of the practice of leasing state convicts to private employers. The final labor plank recommended a national conference of all labor organizations "to discuss such measures as may be of interest to the laboring classes."

The railroad planks betrayed agrarian anxiety over the power of railroad lobbyists to manipulate state legislatures and law enforcement agencies, as well as the ability of railroad financiers to profit from watered stock. Railroad property should be assessed at "full nominal value of the stock on which the railroad seeks to declare dividends." The farmers also demanded an interstate commerce law "to secure the same rates of freight to all persons for the same kind of commodities" and to prevent rebates and pooling arrangements designed "to shut off competition."

The five land planks addressed agrarian grievances that stemmed from the activities, state and national, of Scottish and English cattle syndicates and domestic railroad land syndicates. By 1886 both groups had seriously diminished the remaining public domain available for settlers. The committee demanded that all land held for speculative purposes should be taxed and, in a brusque reference to usurious interest rates, urged that they should be taxed "at such rates as they are offered to purchasers." Also recommended were measures to prevent aliens from speculating in American land and "to force titles already acquired to be relinquished by sale to actual settlers." The committee also demanded that forfeited railroad lands "immediately revert to the government and be declared open for purchase by actual settlers" and that fences be removed, "by force if necessary," from public school lands unlawfully fenced by "cattle companies, syndicates, and every other form or name of corporation." The lone agricultural demand not relating to land policy was one designed to end a capitalist activity that had never found favor with American farmers—"the dealing in futures of all agricultural products."

The most explosive portion of the committee report concerned the financial question. It revealed the extent to which Alliance farmers were being radicalized by the Texas financial community's opposition to the Alliance cooperative movement. The proposal called for a federally administered national banking system embracing a flexible currency, to be achieved through the substitution of legal tender treasury notes for existing issues of private national banks. The sums involved should be issued by the federal treasury and regulated by the Congress to provide "per capita circulation that shall increase as the population and business interests of the country expand." In short, the plank advanced the doctrines of the Greenback Party. To address the immediate problem of a severely contracted money supply, the committee proposed, as a consciously inflationist measure, the "rapid extinguishment of the public debt through immediate unlimited coinage of gold and silver" and "the tendering of same without discrimination to the public creditors of the nation."

Though most Texas Alliancemen were Democrats, the experience of their cooperative movement had persuaded a number of them to make a fundamental shift in their political outlook. The greenback critique of American finance capitalism was not a doctrine one learned in "the party of the fathers." For the farmers, the plank was, in fact, a direct product of their new awareness of the power of the commercial world that opposed the cooperatives of the Alliance.

The seventeenth and final plank was a resolution rather than a demand, but its wording reflected the same concept of the power relationships existing between Alliance members and their state president as had characterized William Lamb's unilateral boycott proclamation six months earlier: the president of the Alliance was "directed" to appoint a committee of three to press the demands on the legislators of the state and nation and "report progress" at the next meeting.

The committee report was the first major document of the agrarian revolt. Its presentation was both electrifying and divisive, and the resulting debate consumed the final two days of the convention. Proceeding behind closed doors, the discussion grew acrimonious as it became clear that a majority of the delegates supported all seventeen demands and were determined to press the committee report to a formal vote. The final tally was 92 delegates in favor of the demands and 75 opposed. By this margin the Alliance had launched a program of political "education"—the necessary foundation for eventual agrarian insurgency. William Lamb's radicalism of January had become, by a narrow margin, the radicalism of the Alliance in August.

11

The "Cleburne Demands" were front page news across Texas on August 8, 1886. The first newspaper reports were restricted to lengthy and precise descriptions of the demands, but as the implications of the Alliance action became clear, newspaper reactions turned hostile. "The Democratic Party is in perilous position," warned the Galveston *News.* The Dallas *News* decided the Alliance had become "dominated by the spirit of class legislation, class aggrandizement, class exclusiveness, and class proscription." The *News* told Democrats that it was "scarcely less than treason to be indifferent to such a danger."

Alliance conservatives also were profoundly disturbed by the Cleburne Demands. Their attitude marked the first surfacing of deeply held cultural presumptions that stood as forbidding barriers to the long-term goals of the People's Party. While Alliance conservatives shared the radicals' concern over the plight of the farmer, they felt, or at least hoped, that they would not have to break with their received political heritage to express that concern effectively. But the eleventh demand of the Cleburne document, the greenback plank, was unacceptable to the Democratic Party, and that fact created an agonizing dilemma for conservative farmers. For some among them, loyalty to the Democrats was a matter of habit and social conformity; for others, it was a pragmatic evaluation of what they took to be the invulnerability of the "party of the fathers" to effective attack from without; for still others, the stance of "nonpartisanship" was simple evidence that their commitment to reform was a step lower on their personal scale of political priorities than an emotional dedication to white supremacy and to its institutional expression in the South, the Democratic Party. Some farmers, finally, simply could not conceive of a Southern farm organization prospering outside the Democratic Party; their loyalty to the party was a function of their loyalty to the Alliance. Whatever the individual variants, the Cleburne conservatives expressed habits of thought that challenged Alliance activists and Populists throughout the years of the agrarian revolt.

After midnight on the evening of the final vote on the Cleburne Demands, a group of conservative "nonpartisans" led by Dunlap met and drafted a public statement of dissociation. Supporters of the demands thereupon drafted a counter-statement, providing details of the tactical maneuvering and upbraiding the minority for publicly revealing divisions within the order. The conservatives then formed a rival "Grand State Farmers Alliance" of an avowedly nonpartisan character. Among other assets, Dunlap's group held all the funds of the regular Alliance, which were safely in the possession of the decamping treasurer. The destructive Lampasas experience of 1879–80 seemed to be repeating itself.

12

Into this sensitive situation stepped one of the most talented and enigmatic men brought forward by the agrarian revolt. In August 1886 Dr. Charles W. Macune was on the verge of an extraordinary career. Born in Wisconsin, and orphaned at ten, he had roamed the West and arrived on the Texas frontier in 1870 at the age of nineteen. He "read" for the professions and in time came to practice both medicine and law. But Macune also possessed untapped talents as an organizer. A strikingly handsome man, he was both a lucid writer and a sonorous, authoritative public speaker. Above all, he was a creative economic theorist. He had joined the order in the winter of 1885–86 and had risen quickly to local prominence in the Milam County Alliance, which elected him as one of its three delegates to the Cleburne convention. Though the 1886 meeting marked Macune's first statewide convention, his farmer associates, impressed with his diplomatic performance during the tense debate over the demands, elected him chairman of the executive committee. He stepped quickly into the vacuum left by Dunlap's resignation and immediately sought to placate the disgruntled conservatives while not incurring the displeasure of the radicals. Aware of the delicacy of the situation, Macune called the order's leadership together and concentrated on two men—Dunlap, spokesman for the "nonpartisan" position, and Evan Jones, spokesman for the radicals. Jones had come into prominence at Cleburne. He was typical of the new brand of local leadership that had emerged in the Alliance since 1883. A "political Allianceman"—that is to say, a greenbacker—he led Erath County farmers into independent political action with a full county ticket against the Democrats in 1886. A vigorous proponent of the Cleburne Demands, Jones, like Macune, had been elected by the Cleburne convention to the state executive committee. Dunlap on the "nonpartisan" right and Jones on the "political" left represented the divergent tendencies in the order—with Macune trying to find tenable ground in between.

Neatly balancing his words, Macune persuaded Dunlap of the "importance and danger" of the conservatives' action. For the moment, the very demonstration of such concern by Macune seemed enough to immobilize Dunlap. He acquiesced in a passive course. To Jones, the advocate of aggressive action, Macune talked of the need for the Texas Alliance to broaden its base by seeking mergers with whatever progressive farm organizations could be found in other states. As Macune doubtless expected, Jones was quite responsive to the suggestion. Only after getting both sides to agree to a special statewide meeting at Waco did Macune finally accept the resignations of Dunlap and the conservative executive committeemen that had been tendered at Cleburne. It can be safely assumed that Macune was able to keep the nonpartisans—and the order's treasury—within the Alliance "family" while shearing them of their power primarily because of the remarkable response of Texas farmers to the Cleburne Demands. The tactical position and the self-confidence of the conservatives had been eroded by the sheer fame of the Cleburne document.

The two Alliance factions were talking to Macune, if not yet to each other. Temporarily, at least, the Alliance was intact—and under leadership that now excluded the order's older traditionalists. An impulse toward aggressive action, greenback in implication but as yet not wholly internalized, had come to power within the Texas Alliance.

As Macune labored behind the scenes, the Alliance grew spectacularly. Farmers by the thousands heard about the Cleburne Demands and decided that the Alliance meant what it said about helping "the industrial classes." Hundreds of charters for new suballiances were issued, and older local groups took on new members until whole farming areas were enrolled. New members were added at a rate of up to 20,000 per month through the autumn and winter of 1886. Alliance lecturers were "sweeping everything before them." By the time Macune convened the special Alliance meeting at Waco in January 1887, the statistics were overwhelming: the Grange numbered less than 9000, while the Alliance claimed 200,000 members in over 3000 chartered suballiances.

13

But the nation had not yet discovered that Alliancemen had more than numbers—that they had, in fact, learned from their experiences a new way of thinking about American economic customs. Alliancemen proclaimed, indignantly, "Shame on your boasted institutions of liberty." They told each other that unless America changed, "you will have only three rights . . . to work, to starve, and to die." Since the American Revolution, the "plain people" of the country had imbibed the cultural teachings generated by the respectable elements of society. Until now, they had, with few exceptions, demonstrated that they had learned good manners and had remained properly deferential to their betters. Where would this new language of politics lead? To a new and more generous democratic community? Or perhaps to revolution and anarchy?

If, by 1886, the nation had not yet learned enough about the Farmers Alliance to worry, Texas conservatives had. The Dallas *News* labored to describe the new Alliance vision:

> The discontented classes are told, and are only too ready for the most part to believe, that the remedy is more class legislation, more government, more paternalism, more State socialism. The current gospel of discontent as a rule is sordid and groveling. Its talk is too much about regulating capital and labor . . . and too little about freeing capital and industry from all needless restraints and so promoting the development and diffusion of a high order of hardy manhood.

In a corporate culture increasingly responsive to the new creed of progress, the Dallas *News* was scarcely alone in its formula for hardy manhood. But the Alliance was exploring another path, one steeped in a new language of cooperation. To those who understood what was being said, the two views of democracy could not easily coexist. If the Alliance grew to national proportions, the country might well have to make a permanent choice. For, among some of the "plain people," a new vision of possibility had come alive in the land.

And—self-evidently—it was not grounded in deference.

3
The Cooperative Vision
Building a Democratic Economy

"the foundation that underlies the whole superstructure. . . ."

To all appearances, the agrarian revolt developed in the South and West in a fashion that merged separate currents of reforming energy. The National Farmers Alliance seemed a powerful tributary of insurgency that conveyed sluggish Southerners toward the People's Party. There the mobilized reform energies of the South flowed into a raging Western torrent. That torrent had no apparent source; it seemed to have materialized from an unknown well-spring concealed somewhere in the Great Plains. The revolt in the West had simply "happened"—times were hard. So the movement of the farmers appeared to the puzzled nation in 1892. Appearances, however, were misleading. The "tributaries" of the People's Party were not divergent; indeed, they were not even tributaries; the radical currents merely needed to be traced to their common headwaters.

The ideological course to the 1892 Omaha Platform of the People's Party ran back through the Ocala Demands of 1890, the St. Louis Platform of 1889, the Dallas Demands of 1888, and the Cleburne Demands of August 1886. For in 1886 the organizational impulses of hope and self-respect generated by farmer cooperatives, impulses that were to lead to the People's Party, identified themselves. Shaped by the tensions of organizing and expanding across the nation, the new culture of a people's politics that had materialized in Texas in 1886 became known to the nation in 1892 as "Populism."

The spectacle of earnest farmer-lecturers setting off on con-
tinent-spanning journeys in the late 1880's to organize the folk,
and, furthermore, doing it, appears now to have had a kind of
rustic grandeur. It was, in fact, the most massive organizing
drive by any citizen institution of nineteenth-century America.
The broader outlines had a similar sweep: the Alliance's five-
year campaign carried lecturers into forty-three states and
territories and touched two million American farm families; it
brought a program and a sense of purpose to Southern farmers
who had neither, and provided an organizational medium for
Westerners who had radical goals but lacked a mass constituency.
Despite the ultimate long-term effects, the massive recruiting
campaign of 1887–91 nevertheless also dramatized the impending tension within the ranks of the new movement that was
beginning to form. The immediate results were far too divisive
to produce any portrait of grandeur, however rustic.

As matters developed late in 1886, the remarkable growth of
the Texas Alliance was a direct product of the accelerating
political momentum within the order. Earnest radicals soon
discovered, however, that this very momentum had produced
an acute internal crisis. After some anxious months the organization preserved its cohesion, but only by a new burst of
creativity keyed to further expansion. More than anything else,
the surge of the Farmers Alliance across the South and then the
West developed out of the order's attempt to save itself from
fratricidal destruction in Texas. This ideological tension had its
origin in a tense internal struggle that took place on the eve of
the campaign to organize the South.

2

Many of the delegates who gathered for Charles Macune's
specially called conference in Waco in January 1887 carried
militant instructions from their home Alliances. The continued
growth of the order since Cleburne had confirmed for the
activists the practicality of aggressive advocacy. Balancing this
thrust was the psychological hold the Democratic Party had on
Southerners. Those who opposed the Cleburne Demands did so

not so much because of their specific content—which faithfully expressed grievances most farmers regarded as legitimate—but because of their implicit repudiation of the Democractic Party of the South.

Amid such contradictory influences, the Waco conference was threatened with partisan discord until, on the second day, Charles Macune offered a transforming proposal. Ignoring the internal divisions, he oriented the delegates toward a more basic purpose—combating the farmer's traditional problem of credit. As Alliancemen of all factions listened attentively, Macune proposed a central statewide "Farmers Alliance Exchange" as a giant cooperative to oversee the marketing of the cotton crops of Alliance members and to serve as the central purchasing medium for Texas farmers. A statewide cooperative was truly a breathtaking vision. But Macune had another challenging proposal as well. He told the delegates that the men who had developed the Alliance cooperative and had organized Texas could "organize the cotton belt of the nation." He outlined his recently conceived plans for projected merger of the Alliance with a small Lousiana farm organization and offered an inspiring vision of a Southwide monopoly of organized agriculture to combat the marketing and financial monopolies of the nation.

The delegates were, to say the least, responsive. Macune brought an individual capacity to act, and a specific plan, at the precise moment when the Alliance had completed its basic organizing job in Texas and was ready—structurally and psychologically—to move to larger tasks. Macune's proposal to organize the nation's cotton belt was a step toward a national presence that every Alliancemen could appreciate, regardless of his political inclinations. Conservative objections to the Cleburne Demands receded in importance in the face of this larger objective. Partisan wrangling was put aside. All seemed relieved that the Alliance, so obviously healthy in other respects, had not been shattered. In a wave of good feeling, the Cleburne Demands were reaffirmed—this time without dissent—and the merger with the Louisiana Farmers Union was approved. A new organization, "The National Farmers Alliance and Cooperative Union," was established. C. W. Macune was unanimously elected

president, and the Louisiana representative who was present
was elected vice-president. Travel funds were allocated, organizers were selected and briefed at a specially called statewide
meeting of lecturers, and the campaign was launched—all within
five weeks of Macune's Waco speech.

3

Six lecturers were initially dispatched to Mississippi, six to
Alabama, seven to Tennessee, five to the border state of Missouri,
and three to Arkansas. Others moved into the Carolinas, Georgia,
Florida, Kentucky, and Kansas. Tested in the process of organizing 200,000 farmers in Texas, they took with them detailed
plans for state characters, county organizations, and suballiances,
and the aggressive anti-monopoly oratory of the Alliance movement. The Macune formula for a centralized buying and selling
cooperative was outlined to gatherings at hundreds of Southern
crossroads from the Gulf to the Ohio River. Farmers were told
how to establish the trade store system on a county level and
members of suballiances were instructed in the advantages of
electing one of their own number as a business agent. All of the
experiences of the founders were drawn upon, as organizers
also explained the value of Alliance cotton yards, Alliance trade
committees, and Alliance county warehouses.

The results were spectacular. Alliances sprouted not only in
every state, but in almost every county of the Old Confederacy.
Dazzled by his success, lecturer J. B. Barry sent in a rhapsodic
report from North Carolina. "In spite of all opposing influences
that could be brought to bear in Wake County, I met the farmers
in public meetings twenty-seven times, and twenty-seven times
they organized. . . . The farmers seem like unto ripe fruit—you
can gather them by a gentle shake of the bush."

The first organizer to leave Texas for the Deep South was the
order's Traveling Lecturer, S. O. Daws. He departed in February
and organized the first Mississippi Alliance on March 3, 1887.
Within six months, thirty Mississippi counties had been organized
and a State Alliance created. Throughout the South, Texas
organizers followed the precedent set by Daws in his 1884 Texas
campaign: the most aggressive local farmers were named as

organizers and briefed in the techniques of conveying the Alliance doctrine. According to a Mississippi historian, "deputies swept to every part of the state," and one even took to announcing his itinerary in the public press! By the end of 1887, twenty-one North Carolinians were at work as Alliance organizers, and the new state leader, L. L. Polk, declared he needed "five times that number" to meet the demand. Alliances had multiplied at such a rate in one North Carolina county that there was "hardly an interval of five miles . . . that does not have an organization."

The pattern was much the same everywhere. The word spread, as it had earlier in Texas, that the Alliance meant what it said about "doing something for the dirt farmer." One Texan organized over 1300 farmers in a single Alabama county. Another stayed an entire year and later claimed to have organized 1500 suballiances in Alabama and Tennessee. But not all farmers were like "ripe fruit." J. M. Perdue, the author of the Cleburne Demands, wrote back from Opelika, Alabama, that many farmers were "so crushed under the crop mortgage system that they have lost almost all hope of bettering their condition." In a comment that reveals the depth of agrarian animosity toward furnishing merchants and the crop lien system, Perdue added that "the threats of the grab-all family have been given out, and many of the poor are afraid to join the Alliance, fearing the major or the colonel will quit issuing rations to them at 50 percent over cash price." In spite of all hazards, however, the organizing momentum of the Alliance gradually conquered whole farming districts, the hesitant tenants joining a bit late, but joining. The demand for lecturers became so great that not enough could be supplied. In more than one county the farmers organized themselves at mass meetings and formally requested organizers to visit them and show them how to establish an Alliance. To baffled Southern Grangers, only metaphors drawn from nature could explain the organizational phenomenon that had engulfed them. The Alliance had "swept across Mississippi like a cyclone." A North Carolinian added that "a great movement, called the Farmers Alliance, has about ruined the Grange."

By the time the National Farmers Alliance and Cooperation Union of America held its inaugural convention at Shreveport, Louisiana, in October 1887, the cooperative campaign had invaded ten states—all in eight months. President Macune announced that the Texas organizers had brought some $2,866.50 in dues money to the national treasurer, more than enough to repay a $500 loan advanced by the Texas Alliance. After a generation of unrelenting poverty, the farmers of the South were desperate for a message of economic salvation. The Southern yeomanry had indeed become "like ripe fruit." But the memories of political and social orthodoxy were deeply imbedded. The question posed itself: ripe for what?

4

Throughout the Western plains, the question posing itself was of a different order at first. The succession of third party defeats throughout the 1880's in the West seemed, to many radicals, to doom any possibility of broad-based reform. The swift death of the Union Labor Party in 1888 so disillusioned one young radical in Kansas that he retreated to a self-conscious display of bravado in the face of what he clearly regarded as a hopeless situation:

> I know that for the man who sees the evils of the time—the want, ignorance and misery caused by injus laws—who sets himself so far as he has strength to right them, there is nothing in store but ridicule and abuse. The bitterest thought, and the hardest to bear, is the hopelessness of the struggle, "the futility of the sacrifice."

The author of these somber words was Jeremiah Simpson, a man destined within a very short time to achieve national notoriety as "Sockless Jerry" Simpson, the very symbol of fiery prairie Populism. Indeed, a mere eleven months after he brooded over ranks grown "thin by death and desertion," Simpson offered a remarkably different appraisal of the health of the reform movement: "Our meetings are growing; at first they were held in country school houses while the other parties held theirs in the open air; now ours are outside, and the other parties are never heard of at all."

To reform candidates who saw politics solely in terms of the size of crowds at summer speech festivals, a marvelous change had indeed come over the people of Kansas sometime between 1889 and 1890. But if the gulf between Simpson's exaggerated despair one year and his romantic optimism the next had the merit of pointing to anything at all of substance, it was that essential forms of social change are rarely those most easily seen from the rostrums of meeting halls.

The organizing problem facing reformers in the West was not a result of some rare flowering of agricultural prosperity. During the late 1880's when Simpson and others labored against the "evils of the time," agrarian distress wracked Kansas almost as fully as it did the South. But as Simpson clearly failed to grasp, the organizing hazard did not revolve simply around economic conditions. Indeed, coherent reform politics never does. Insurgent movements are not the product of "hard times"; they are the product of insurgent cultures. "Hard times" demoralize people, making coherent politics even more difficult than normal. A generation on the crop lien had taught millions the power of this political fact. Effective insurgent cultures, on the other hand, offer people hope. And from this starting point, political movements are possible—especially if times are hard. What came to Jerry Simpson's region of Kansas in 1889 was the political culture of the Alliance cooperative movement. The cooperative vision was carried to the farthest reaches of the state by scores of lecturers. In due course, a political earthquake resulted.

In charting the agrarian revolt in Kansas, therefore, it is necessary to start not with reform politicians like Simpson (who often give voice to insurgent impulses they do not fully understand), but with the men who helped generate the movement itself. The S. O. Daws of Kansas was a young man in his mid-twenties named Henry Vincent.

5

Some months before the Texas Alliance deployed its lecturers through the South, the town of Winfield, Kansas, in Cowley County near the Oklahoma border, discovered that it had suddenly acquired a young radical who had descended upon the quiet hamlet with the intention of opening a newspaper.

That young man was to prove one of the most energetic Populists of them all.

Henry Vincent was one of five sons of James Vincent, a rather unusual man in his own right. A radical egalitarian, abolitionist, and free thinker, the senior Vincent put himself through Oberlin College as a youth, married, and moved to a farm near Tabor, Iowa. He became western correspondent for Greeley's *New-York Tribune* and Garrison's *Liberator*, worked for the American Anti-Slavery Society, and operated an underground railroad and school at Tabor. He seems to have rather thoroughly transmitted his humanist views and a measure of his tenacity to all of his children—especially to young Henry. At the age of seventeen, Henry began a journal called the *American Nonconformist*, which he printed on a thirty-five-dollar hand press in Tabor. Seven years later, he, along with two of his brothers, thirty-year-old Cuthbert and twenty-three-year-old Leopold, migrated to Cowley County in Kansas. The brothers selected Winfield for the site of their journalistic effort after a careful inspection of other locales in what seems to have been a search for a climate conducive to social change. Grandly entitling the venture *The American Nonconformist and Kansas Industrial Liberator,* Henry and his brothers unleashed among their boom-conscious Kansas neighbors a startling brand of journalism. The inaugural issue of October 7, 1886, declared:

> This journal will aim to publish such matter as will tend to the education of the laboring classes, the farmers and the producer, and in every struggle it will endeavor to take the side of the oppressed as against the oppressor, provided the "underdog" has concern for his own hide to defend himself when he is given the opportunity, and not turn and bite the hand of him who has labored for his freedom, by voting both into a worse condition than before.

The Vincents established a far-flung system of exchanges with other agrarian and labor newspapers and kept a careful nation-wide watch on political activities in behalf of the "underdog." They soon became well-informed on the activities of both the Knights of Labor, then at its national peak of influence, and the

Farmers Alliance, then still confined to its Texas origins. The *Nonconformist* praised the racial liberalism demonstrated by the Knights at their integrated 1886 convention in Richmond, Virginia, and similarly applauded the aggressive spirit of the Cleburne Demands, which emerged from the little-known farm order in Texas in the same year. It duly noted the spread of the Alliance across the South, and when the Texas organizers ranged over Missouri and edged into neighboring Kansas in the summer of 1887, they quite naturally found a welcome first and foremost in Cowley County. One Texan organized a number of suballiances in the county and a second arrived later in the year to extend the Alliance message to surrounding counties. The Vincents meanwhile were busy applauding the efforts of Western radicals to reconstitute a new national third party as a home for displaced greenbackers. In the course of these third party efforts, the Vincents established relations with the radicals in the Texas Alliance who had engendered the Cleburne Demands.

By the time the third party men were ready to hold a nominating convention for the new national Union Labor Party in Cincinnati in the spring of 1888, a coterie of politically insurgent Westerners had established contact and had begun working together. The network, which was primarily composed of Kansans and Texans, also included greenback veterans from Iowa, Alabama, Nebraska, Arkansas, Missouri, and the Dakotas.

While the subsequent poor showing of the Union Labor Party in the fall elections of 1888 helped plunge radical politicians such as Jerry Simpson into despair, there was a more germane development—the spread of the Kansas Farmers Alliance from its base in Cowley County. The enormous growth of the order in Texas in 1885–86 came only after the founders had earlier consolidated a strong geographical base in a nucleus of counties, in the process perfecting their organizational doctrines and an accompanying rhetoric of recruitment. In 1888, the Kansas Alliance consolidated a similar base in south-central Kansas. Henry Vincent's effort to familiarize Kansas farmers with Alliance doctrines was not simply one of journalistic "education." Vincent understood that movements need organization and organizations need internal lines of communication. Late in

1888, following the disastrous third party showing in the November elections, Vincent led a group of Kansas radicals on a pilgrimage to Dallas for further briefings on the structure of the Alliance movement at the organization's national offices. Additionally, W. P. Brush, in the course of his official duties as state organizer for the Kansas Alliance, attended the 1888 convention of the National Farmers Alliance at Meridian, Mississippi. There he acquired further details of the cooperative movement and was appointed one of twelve "national lecturers" designated by the Macune-led Southerners.

In December 1888 the Farmers State Alliance of Kansas was organized. Lecturers armed with charters and equipped with the new lecturing methods spread over the state. They met with an immediate response, just as their counterparts in the South had earlier. Kansas farmers listened to the outline of the cooperative program and promptly joined the Alliance in droves. Suballiances elected business agents and county Alliances formed trade committees. As evidence of the success of the organizing campaign mounted, the *Nonconformist* interviewed Ben Clover, the new Kansas Alliance president on the progress of the order. He announced that "extensive movements are on foot." It was not an idle remark. The new culture of politics was being born. As the cooperative banner went up in central and eastern Kansas, a new and surprisingly democratic energy began to surface in ways that newspapers, friendly and otherwise, could scarcely ignore. In February 1889, the Harper County Alliance began bombarding the state legislature with appeals for stricter usury laws, and two months later Brown County farmers staged a mass protest against what they called "the extortions of the binding twine trust." The farmers decided to "proceed at once to the erection of a cooperative manufactory for binding twine." Elsewhere throughout Kansas, Alliance country trade committees met with merchants, wholesalers, and manufacturers. Where the farmers were rewarded with small successes, their respect for their own efforts and for their Alliance grew. When they encountered rebuffs, they talked about it together and debated the meaning of existing commercial relationships in American society. They also learned that the voting records of their politicians indicated a surprising responsiveness to the needs of banks, railroads, and other corporations. These discoveries did not fit with the description of the Republican Party of Kansas that they had been taught since infancy to believe. Some of the farmers began to get more angry the more they learned, and all of them wanted to do something about it. The suballiance in Kansas, like its predecessors in Texas and the South, became a schoolroom of self-education.

This process took time. Political results first became visible in the handful of "old" Alliance counties near the Kansas-Oklahoma border that had first heard the message of the Alliance in 1887. The 1889 elections in Henry Vincent's Cowley County brought a hint of storms soon to come. Area farmers, already organized as an Alliance cooperative, further organized themselves in a mass meeting and nominated a complete independent ticket against the Republican incumbents.* Affiliating with neither major party, the ticket was elected by a surprisingly large margin. The political culture of the agrarian movement had literally reached a majority of the people in the county.

In steadfastly Republican Kansas, this result was a bit startling in itself. A glance behind the returns, however, yielded even more instructive insights. Whatever had happened in Cowley County over the preceding twelve months, it had clearly not happened countywide. For example, the townsfolk of Winfield had voted Republican in 1888 by a comfortable eight to five margin and had even added slightly to the party's plurality the following year. But a phenomenal change had come over the rural districts of the county. The farmers who voted at Rock Creek precinct were typical. In 1888 they had cast 96 votes for

---

* While the internal dynamics guiding the development of insurgent demo-cratic movements are not generally well understood, the greatest confusion concerns not the building of movements, but rather the process of their politicization. Insurgent economic movements, for example, do not "inexorably" move into insurgent politics. The reason is elementary: the great mass of participants have older cultural memories which shape political conduct along traditional lines and make insurgent politics difficult for the average citizen to imagine. For political insurgency to occur on a mass scale, movements, once recruited, need to be politicized, which is a complex process. The politicization of the mass of Alliance farmers—a development that defined the extent and the limits of American Populism—is traced on pp. 84–87, 91–93, and 125–82.

the Republican Party, 23 for the Union Labor Party, and 60 for the Democrats. In the following year, however, the Republican vote plummeted to 45 while the new independent ticket swept the precinct with 117 votes. Clearly, something was happening at the forks of the creek.

What had happened, of course, was the cooperative crusade. The cornerstone of the independent political movement in Cowley County was a new enterprise created by the Alliance and known as "The Winfield Co-operative Mercantile and Manufacturing Association." Virtually every farmer in the county was a member. A founding trustee of the Alliance cooperative—the first in Kansas—was the editor of the *Nonconformist*, Henry Vincent.

6

What was true for Kansas and Texas was true everywhere; indeed, the agrarian revolt cannot be understood outside the framework of the cooperative crusade that was its source. Amidst a national political system in which the mass constituencies of both major parties were fashioned out of the sectional loyalties of the Civil War, the cooperative movement became the recruiting vehicle through which huge numbers of farmers in the South and West learned to think about a new kind of democratic possibility in America. The central educational tool of the Farmers Alliance was the cooperative experiment itself. The massive effort at agrarian self-help, and the opposition it stimulated from furnishing merchants, wholesale houses, cotton buyers, and bankers in the South and from grain elevator companies, railroads, land companies, livestock commission agencies, and bankers in the West, brought home to hundreds of thousands of American farmers new insights into their relationship with the commercial elements of American society. Reduced to its essentials, the cooperative movement recruited the farmers to the Alliance in the period 1887–91, and the resulting cooperative experience educated enough of them to make independent political action a potential reality. In the process, the Alliance created the world's first large-scale working class cooperative and proposed a comprehensive democratic monetary system for America, the world's emerging industrial leader. That the chief theorist of both the cooperative and the monetary system, Charles Macune, consistently opposed Alliance political activism and feared the emergence of the third party added a curious dimension to the internal politics of the agrarian revolt.

7

Irony dogged Charles Macune throughout his years as national spokesman of the Farmers Alliance. The scores of Alliance organizers who fanned out across the South and Midwest in 1887–89 and who sallied north of the Ohio River and across both the Rockies and the Appalachians in 1890–91 had been defined by their previous experience not only as "lecturers," but also as incipient political radicals. Their duties as lecturers carried them to the remotest backwaters of the rural countryside and into the very maw of the crop lien system of the South. Upon the lecturers fell the burden of explaining the function of the suballiance business agent, the county trade committee, and the visionary state exchange that, in the South, might free them all from the furnishing merchant. Upon the same lecturers also fell the burden of explaining the delays, the opposition of merchants, bankers, commission agents, and sundry other functionaries who came to represent "the town clique." Whether in Alabama or South Dakota, a cooperative encountering difficulty constituted an implied rebuke to the Alliance leaders who had praised the idea in the first place. At such a time of difficulty—and it came, eventually, to every Alliance cooperative in every state, from Florida to Oregon—the farmers, so recently brought to a new level of hope by the promise of their movement, looked to the messengers of cooperation both for explanation and guidance. The latter responded in one of two ways. They could blame the difficulties on the farmers, assert that rank and file members did not understand cooperation well enough, and insist that they expected too much, too soon. In the course of this response they could counsel patience and devote much time to thorough explanations of the theory and practice of cooperation. Many Alliance leaders followed such a course, none with more grace than Macune himself. But the mass program of agrarian cooperation that Macune had visualized also set in motion another and rather different response. In this view, cooperative difficulties were not inherently the fault of the idea itself, nor were they traceable to deficiencies in farmers generally; rather, cooperatives encountered trouble because of the implacable hostility of the financial and commercial world. Alliance leaders holding such views could explain that town merchants selectively cut prices to make the Alliance trade store look bad and that bankers refused to take the notes of the Alliance state exchange because the bank's mercantile clients wanted the exchange to go under. They could add that the Alliance store or warehouse often performed its duty even if it sold not a dollar of goods—if, simply by its presence, it introduced genuine wholesale and retail competition into rural America. Macune was also capable of this analysis, and sometimes with a passion that rivaled the anti-monopoly intensity of spokesmen for the Alliance's left wing.

But however individual leaders responded initially, the cooperative crusade and the experiences it generated set in motion an intense dialogue within the Alliance, one that reached realms of radical political analysis that Macune had not foreseen and from which he instinctively recoiled. Yet the cooperative experience increasingly set the terms of the debate in ways neither the order's moderates nor its radicals could conveniently ignore. The mass of farmers themselves saw to this. They wanted freedom from what they regarded as intolerable conditions, and the delays they experienced in their local and state cooperative efforts brought forth questions to which lecturers had to respond, whatever their politics. Indeed, a raw irritant persisted at the core of this internal Alliance dialogue between the farmers and their spokesmen, a criticism by the rank and file that, however respectfully implied, ate at the very heart of the lecturers' sense of justice. It was not the lecturers' fault! They felt, more tellingly than some of them could explain, the totality of banker and merchant opposition, the calumny of news stories about the Alliance in the metropolitan press, the deceptive style of traditional politicians who voted for the commercial classes while pretending to be friends of the people. In ways that Macune both understood and resisted, Alliance leaders came to have such thoughts in the late 1880's, and their knowledge constituted an organizational imperative for a new political party free of the control of bankers and their allies. The men who believed in cooperation the most, the Alliancemen who became trustees of local trade stores, county warehouses, and state exchanges, who counseled against politics as a divisive influence at a time when the cooperative needed the support of a united farming class, who then saw the cooperatives stagger under financial burdens born of lack of credit—it was some of these very men, Macune's lecturers, who eventually carried the Farmers Alliance to the People's Party. The interior logic of the cooperative crusade, both in the hopeful early days of recruitment and in the more difficult days of implementation, drove Alliance leaders toward such ultimate political choices because, quite simply, the structure of American society impelled it.

8

The reality that explained the remarkable organizing potential of the Alliance cooperative rested in the substance of the daily lives of millions of farmers. In the late nineteenth century a national pattern of emerging banker-debtor relationships and corporation-citizen relationships began to shape the lives of millions of Americans. Throughout the Western granary the increasing centralization of economic life fastened upon prairie farmers new modes of degradation that, if not as abjectly humiliating as Southern forms, were scarcely less pervasive.

Of the many ingredients in the new way of life, most easily understood is the simple fact that farm prices continued to fall year after year, decade after decade. The dollar-a-bushel wheat of 1870 brought 80 cents in 1885 and 60 cents in the 1890's. These were official government figures, computed at year's end when prices were measurably higher than those received by the farmer at harvest. Actually, most Dakota farmers received closer to 35 cents a bushel for their wheat in the days of Populism. Similarly, the 1870 corn crop averaged 45 cents a bushel;

thereafter it fell steadily, plunging below 30 cents in the 1890's, according to official figures. But as early as 1889 corn in Kansas sold at 10 cents a bushel, the U.S. Agriculture Department's figures to the contrary notwithstanding.

Moreover, the grain that made the nation's bread was a demanding crop; it had to be harvested at breakneck speed each fall before it became too dry and brittle to bind well. The Midwestern farmer went into debt to buy the needed equipment. He made chattel mortgage payments on such machines at rates of annual interest that ranged from 18 to 36 per cent and in currency that appreciated in value every year. Under these circumstances, the steady decline of commodity prices further reduced his margin of economic maneuver.

But this was only a part of the problem facing Western agriculturalists. Like most nineteenth-century Americans, farmers were enthusiastic about the arrival of each new railroad that promised to further "open up the country" to new towns and new markets. But the farmers' euphoria at the appearance of a new rail line inevitably turned to bitter resentment. The farmer in the West felt that something was wrong with a system that made him pay a bushel of corn in freight costs for every bushel he shipped—especially since the system somehow also made it possible for large elevator companies to transport grain from Chicago all the way to England for less money than it cost a Dakota farmer to send his wheat to the grain mills in near-by Minneapolis. A number of railroads also forced shippers—grain dealers as well as individual farmers—to pay freight costs equal to the rail line's most distant terminal, even should they wish to ship only over a lesser portion of the line. (The extra fee was called "transit.") In many locales, the farmer had the option of paying or seeing his crop rot.

Underlying the entire new structure of commerce was the national banking system, rooted in the gold standard and dominated by Eastern commercial banks, most prominently the House of Morgan. Though bankers profited from the high interest rates that accompanied tight money, the gold-based currency was so constricted that virtually every year the calls on the Eastern money market by Western banks at harvest brought the nation dangerously close to financial panics. The sheer burden of providing the necessary short-term credit to finance the purchasing and shipping of the nation's annual agricultural production was more than the monetary structure could stand. The prevailing system, however, did have the effect of administering a strong downward pressure upon commodity prices at harvest time.

On the "sod-house frontier" of the West, the human costs were enormous. Poverty was a "badge of honor which decorated all." Men and children "habitually" went barefoot in summer and in winter wore rags wrapped around their feet. A sod house was a home literally constructed out of prairie sod that was cut, sun-dried, and used as a kind of brick. Out of such materials, the very "civic culture" of the agrarian revolt was constructed. The farmers knew where their problems were. A social historian reports that farmers in the 1870's and 1880's reserved their "deepest enmity" for grain and stock buyers and for the railroads that served farmers and middlemen alike.

Everywhere the farmer turned he seemed to be the victim of rules that somehow always worked to the advantage of the biggest business and financial concerns that touched his world. To be efficient, the farmer had to have tools and livestock that cost him forbidding rates of interest. When he sold, he got the price offered by terminal grain elevator companies. To get his produce there, he paid high rates of freight. If he tried to sell to different grain dealers, or elevator companies, or livestock commission agents, he often encountered the practical evidence of secret agreements between agricultural middlemen and trunk line railroads. The Northern Pacific named specific grain terminals to which farmers should ship, the trunk line simply refusing to provide railroad cars for the uncooperative.

Among the large new business combinations that were engaged in creating trusts in virtually every major branch of American commerce, the watering of stock became so routine that it is not too much to say that the custom provided the operating basis for the entire industrial system that was energing. To pay even nominal dividends on watered stock, companies needed high rates of profit—in effect, they converted their customers into

real sources of direct capital. Railroad networks that cost
$250,000 in public money to build were owned by companies
that capitalized themselves at $500,000 and then sold construc-
tion bonds on $500,000 more. Agrarian spokesmen wondered
out loud why the citizenry should be paying interest on public
indebtedness of one million dollars, 75 per cent of which was
watered stock. Railroad magnates rarely bothered to reply to
such critics beyond offering an occasional opinion that popular
concern about the interior affairs of business corporations was
"officious."

9

But widespread as suffering was throughout the West, nowhere
in America did the burdens of poverty fall more heavily than
upon the farm families of the rural South. The crop lien system
had driven millions west in the 1870's; by the late 1880's the
system had graduated to new plateaus of exploitation: as every
passing year forced additional thousands of Southern farmers
into foreclosure and thence into the world of landless tenantry,
the furnishing merchants came to acquire title to increasing
portions of the Southern countryside. Furnishing men had so
many farms, and so many tenants to work them, that it became
psychologically convenient to depersonalize the language of
agricultural production. Advancing merchants spoke to one
another about "running 100 plows this year," a crisp phrase that
not only referred to thousands of acres of land but also to
hundreds of men, women, and children who lived in peonage.
Through a 1500-mile swath of the Southland, from Virginia to
Texas, such absence of choices came to characterize the monot-
onous lives of millions, tenants and "landowners" alike. It was
into this vast domain of silent suffering that the lecturers of the
Farmers Alliance deployed in 1887–88. The results were difficult
to describe, though a South Carolina Granger made an effort by
reporting to his national offices that the Alliance had "swept
over our state like a wave."

The message of the lecturers was persuasive because the
goal—to change the way most Southerners lived—was one the
Grange had never dared to attempt. The larger dream contained

in the new and untested Texas plan for a central state exchange added the final galvanizing ingredient to the formula of hope that was the Farmers Alliance.

But the new recruits did not wait idly for the arrival of a statewide marketing system. Hope was everywhere and across the South, farmers in a dozen states competed with one another in pioneering new varieties of purchasing cooperatives that could be constructed to defeat money-lenders and wholesale and retail merchants. The leader of a local Georgia Alliance wrote the order's national journal in Texas that "we are going to get out of debt and be free and independent people once more. Mr. Editor, we Georgia people are in earnest about this thing." The effort "to be a free and independent people" was emphasized by North Carolina Alliance leader L. L. Polk, who explained: "There is a limit, even to the submissiveness of farmers."

Just how high the stakes were, and how far from "submissiveness" the farmers had to carry themselves, was brought to light by the cooperative movement in Alabama. After Alliance-men in Dothan had cooperatively erected a warehouse and had begun to demonstrate its utility by instituting cooperative pur-chasing and marketing arrangements, merchants, bankers, and warehousemen succeeded in inducing the town council to levy a $50 tax on the warehouse. The farmers responded by moving their building outside the city limits, whereupon the council attempted to make them pay for draying their cotton into and out of town. Attempts to enforce the ordinance led to a gunfight in which two men were killed and another wounded. Alabama farmers thought these events over for awhile, and it was not long before the Alliance began to explore ways to start a cooperative bank. Though the means to carry out their proposal were never found, it was clear that the cooperative experience provided remarkable lessons in how power worked in America.

Yet for all the hope that it stirred, the Alliance movement remained just that—a hope. Though the farmers came by the thousands, the Alliance found it difficult to implement an effective cooperative program. While local arrangements might produce marginal improvements, a statewide undertaking was needed to cope with the lien system. As Southern Alliances grew, and as both farmers and their spokesmen learned more about the intricacies of cooperation, they increasingly tended to keep one eye cocked on Macune's Texas Exchange, where the mass marketing concept was undergoing its first major test in the South. In 1887–88 the Texas effort entered its crucial stage— and Alliance leaders throughout the nation watched in hope and apprehension as the idea of a cooperative commonwealth endeavored to coexist with banker-centered American capitalism.

10

The original plan of a central state exchange which Macune had outlined in January 1887 had been consummated in stages. Macune learned that a central exchange in Texas could sell directly to Eastern factories if it possessed sufficient capital to underwrite its contracts properly. He reported this to his Texas colleagues, and the Alliance, after taking competing bids from several cities, selected Dallas as the site for the exchange, receiving a bonus of $3500. On this shred of capital the "Farmers Alliance Exchange of Texas" opened for business in Dallas in September 1887. Macune used the internal organizational structure of the Alliance to bring the individual farmer closer to the prevailing prices being paid on the world cotton market. Each county business agent, working through the Alliance cotton yard established by each county trade committee, was instructed to weigh, sample, and number the bales of cotton in his local yard. The local agents wrapped samples from each bale, placed tickets on them giving weight, grade of cotton, and yard number, then expressed them in sacks to the Dallas exchange. There they were placed on display in a sample room where cotton buyers could examine them. Export buyers were impressed, and they came to the Texas Exchange. In one massive transaction, 1500 bales of cotton were selected from samples in Dallas and sold for shipment to England, France, and Germany. The cotton was shipped from twenty-two different stations in Texas.

Impressive as all this activity was as a demonstration of agrarian self-help, the Texas Exchange leadership soon realized it was not nearly enough, for it did not free the small farmer or the tenant farmer from all the personal and financial indignities involved in the crop lien. The directors soon found themselves awash in petitions from suballiances calling for the implementation of some plan so they could "make a crop independent of the merchant."

Late in 1887 the directors of the Texas Exchange faced the challenge squarely. The plan they embarked upon was one of the most creative in the annals of American farm organizations, and it led directly to the one pathbreaking political concept of the agrarian revolt, the sub-treasury system. In November 1887, the Texas Exchange advanced its program. It was called the "joint-note plan." The quiet phrase concealed a breathtaking extention of the cooperative concept: landowning farmers in the Alliance were asked to place their entire individual holdings at the disposal of the group—to stake their own futures on the ultimate success of the Alliance cooperative. The gamble was an unusual one: landowners and tenants alike would collectively purchase their supplies for the year through the state exchange on credit, the landowners signing the joint note. For collateral, they would put up their land and endeavor to protect themselves against loss by taking mortgages on the crops of the tenants. They would market their cotton collectively through the exchange and then pay off the joint notes at year's end. The farmers would sink or swim together; the landless would escape the crop lien, too, or none of them would. As they had in the past, the brotherhood of the Alliance would "stand united." In one dramatic season of cooperative marketing and purchasing, they would collectively overcome all of the furnishing merchants of Texas and free every farmer in the state from the clutches of the lien system!

The planning behind the joint-note cooperative mobilized the entire infrastructure of self-help that the order had created through the years. Each Alliance county business agent was to acquire from each suballiance member who wanted supplies on credit a schedule of his probable individual needs for the coming year, together with a showing of "full financial responsibility" and a pledge of cotton worth at least three times as much as the amount of credit requested. The farmers of each suballiance were then to execute a collective joint note for the estimated amount of supplies for all of them. The note was to draw interest after May 31 and was to be paid November 15, after the 1888 crop had been harvested. Each signer of the joint note was required to specify the number of acres of land he owned, its value, outstanding indebtedness, the number of acres cultivated both in cotton and grain, and the value of his livestock. He also agreed to allow his cosigners to harvest his crop in the event he became incapacitated. Such joint notes from suballiances were then to be screened by the county business agent and the county trade committee before being passed on—along with the collective supply order of all participating suballiance members—to the "committee of acceptance" in the state exchange in Dallas.

The third part of the plan was, of course, that the exchange would use the notes as collateral to borrow money to purchase the supplies. The supplies were to be shipped on a monthly basis, one-sixth of the order for each of the months between May and November. The notes would draw 1 per cent per month from date of shipment until the individual farmer repaid his debt. In addition to making substantial savings in credit costs, (5½% compared to 50 to 200% charged by furnishing merchants) farmers might expect much cheaper prices on supplies as a result of the exchange's bulk purchasing power.

To explain this elaborate plan the Texas Alliance mobilized its lecturing system once again, utilizing many of the men who had been seasoned in the organizing campaign in the South the year before. The procedure and all its complicated components were outlined to every county Alliance in Texas and eventually to every suballiance. The farmers marshalled all of their resources in a gigantic effort "to become a free and independent people."

As the joint notes flooded into the Dallas nerve center, efforts to induce the membership to pay the $2.00 per capita assessed for the capital stock of the exchange were intensified. When the exchange directors met in March 1888 to review the situation they found that, while paid-up stock had climbed from its September value—virtually nothing—to $17,000, this "blood money" from the farmers of Texas had to be balanced against the steadily rising amounts contained in the joint notes approved by the committee of acceptance. The committee, anticipating brokering the joint notes through bank loans, began buying supplies to fulfill the orders from the suballiances. By April, when the paid-in capital stock had climbed above $20,000, the joint notes approved by the exchange totaled no less than $200,974.88 on local collateral of over $600,000 in the land and stock of Texas farmers. On this foundation, the exchange had ordered goods in the amount of $108,371.06.

Macune had meanwhile begun the quest for outside banking capital in March, using the joint notes as collateral. After repeated conferences in Dallas, bankers there refused to advance loans. Macune then went to Houston, Fort Worth, Galveston and New Orleans. In Houston he acquired one loan of $6000 by pledging $20,000 in joint notes as collateral, and some mercantile houses advanced supplies, taking the notes as security, but with few exceptions the answer elsewhere was "no." Macune's regular report to the exchange directors sounded an ominous note:

> The business manager spent the whole of the month of March in trying to negotiate banking arrangements whereby a loan could be affected at a reasonable rate of interest . . . but all the efforts made were unsuccessful, and tended to produce the conviction that those who controlled the moneyed institutions of the state either did not choose to do business with us, or they feared the ill will of a certain class of business men who considered their interests antagonistic to those of our order and corporation. At any rate, be the causes what they may, the effort to borrow money in a sufficient quantity failed.

The Texas exchange was suddenly in serious trouble. Obligations assumed by the exchange for supplies shipped to suballiances fell due in May, and the agency was unable to meet them. The Alliance leadership rallied to the exchange. Evan Jones, the Erath County radical now elevated to state president, presided over a joint meeting of the order's executive committee and the exchange's directors. Finding the exchange's books in order, Jones confirmed to his members their hope that "the entire business is and has been conducted upon sound, con-

servative, practical business principles." He added grimly: "It is
time for each brother to realize that faltering now means
unconditional surrender." Jones and the exchange directors
then announced that "grave and important issues confront us.
. . . Unjust combinations seek to throttle our lawful and legitimate
efforts." In order that "proof of the existence of this combination"
could be submitted to the membership, the Alliance leaders
"most earnestly recommended" that a mass meeting be held at
the courthouse in each county of the state on the second Saturday
in June. At these meetings documentary evidence "disclosing
facts of vast importance" would be submitted along with a plan
to meet the crisis.

11

"The day to save the exchange . . ."

That "second Saturday in June" fell on June 9. It was a day men
remembered for the rest of their lives. The response to the call
made one fact dramatically clear: the central reality in the lives
of most Southern farmers in the late nineteenth century was the
desire to escape the crop lien.

The farmers came by the thousands to almost 200 Texas
courthouses on June 9, 1888. Townspeople in the far reaches
of the state who were not privy to the Alliance internal communication network of circulars, lecturers, or the columns of
the *Southern Mercury* knew little of the "Farmers Exchange" and
its financial troubles in Dallas. They watched with puzzlement
the masses "of rugged honest faces" that materialized out of the
countryside and appeared at the courthouses of the state. In
scores of county seats, the crush of farm wagons extended for
blocks in every direction, some beyond the town limits. Reporters
remarked about the "earnestness" of the effort to save the
exchange and the "grim determined farmers" who were making
pledges of support to some far-off mercantile house. The
correspondent of the Austin *Weekly Statesman* recorded that
observers were "completely astonished by the mammoth proportions" of the turnout. In the far north of Texas, 2000 farmers
from over 100 suballiances in Fannin County marched impas-

sively down the main street of Bonham behind banners that proclaimed "The Southern Exchange Shall Stand." A brass band led the procession to the meeting place where Alliancemen stood for hours in the summer heat to learn about the plight of their exchange.

Little in the farmers' experience led them to doubt the interpretation of recent events made by their state officers, for they had been exploited for years. They reported soberly in their county Alliance journals that "we found the trouble was that a number if not all of the banks and wholesale merchants in Texas had turned against the exchange." In farmhouses all over Texas women dug into domestic hideaways for the coins that represented a family's investment in the hope of escaping the crop lien. At a mass meeting in southeast Texas frugal German farmers collected $637.70 and one of their number respectfully tendered a five-year lease on some property to the state exchange for use as it saw fit. In Parker County in northwest Texas the crowd of farmers was so large the hall could not accommodate it and the mass meeting moved in a body to another location. In the county seat of Rockwall, near Dallas, a "large and enthusiastic mass meeting" lasted from 9:00 a.m. to 6:00 p.m., and, though the reporter was hazy about the specifics, it was clear, he said, that they "intend to stand by the exchange." The Hays County Alliance wired that it "loves Dr. Macune for the enemies he has made," and Lamar Alliancemen paid their respects to newspaper rumors about Macune's malfeasance by tending their "thanks and appreciation to the manager of the state exchange." County presidents sent wires to each other as if the sense of solidarity thus engendered might produce greater contributions from a membership notoriously low in capital assets. An East Texas Alliance received telegrams from seven County Alliances, some halfway across the state. A North Texas County contributed $4000 to the exchange, while Gulf Coast Alliancemen pledged several thousand more before adjourning to a joint meeting with the Knights of Labor to plan a local political ticket. On the old frontier, other gatherings, including a well-attended mass meeting in drought-wracked Young County, pledged smaller amounts, and in the center of the state

Bell County farmers announced that their pledges constituted "a success in every respect."

Following this remarkable demonstration from the grass roots, Macune announced that results "had exceeded expectations" and that the exchange was on solid footing. But though pledges seemed to have totaled well over $200,000, the central truth of the Texas Alliance was that it represented precisely those it said it did—the great mass of the agricultural poor. A letter to the *Southwest Mercury* earlier in the year portended the outcome of the June 9 effort: "We voted the $2.00 assessment for the exchange, and as soon as we are able, will pay it, [but] we are not able to do so at present." The dignity with which the admission was made, and the willingness of the *Mercury* to eschew proper "promotional" techniques by printing it without comment, is indicative of the commonplace recognition of prevailing poverty among great numbers of those intimately associated with the Alliance movement, landowners and tenants alike.

In the light of existing economic conditions in the farming districts, June 9 was indeed a success, but not in the scale of six figures. The Texas Exchange eventually received something over $80,000 from its feverish effort. The farmers' pledges represented their hope for the exchange; their actual contributions measured the reality of their means.

12

The exchange survived the season, though manifestly it was not well. Credit—the farmer's age-old problem—was the exchange's problem too. In Dallas, during the summer of 1888, Macune put his fertile mind to work on this old truth that had been so forcibly reaffirmed to him. Somehow, the farmer's crop that the advancing merchants took as collateral had to be utilized by the farmer to obtain credit directly. Precisely how this could be done obviously required something less ephemeral than joint notes that bankers would not honor. But what?

The August 1888 convention of the Texas Alliance provided a tentative answer. Though the Texans were now at the peak of their organizational strength and political influence in the national agrarian movement, the delegates, representing some 250,000 Texas Alliancemen, were absorbed in only one issue, the future of their cooperative exchange. Macune faced an anxious gathering of Alliance county presidents and lecturers on the convention's first day. His exhaustive report, running to sixty-six typewritten pages, presented in minutest detail the record of his stewardship. The stakes were high and all knew it. "Not a few there were," said the Dallas *News*, "who had come many a mile to hear this report and who believed that on this document depended in a large degree the rise or fall, the success or destruction of this experiment." Macune needed all his persuasive wiles, for his address went beyond a mere defense of his direction of the exchange. It included an entirely new approach to the central management of agricultural credit. His months of intensive brooding about the causes of his own and the exchange's crisis had produced a bold new stratagem: he intended to escape the need for bank credit by generating the necessary liquid capital from the farmers themselves.

Macune proposed the creation of a treasury within the exchange to issue its own currency—exchange treasury notes—in payment of up to 90 per cent of the current market value of commodities. Farmers would circulate these notes within the order by using them to purchase their supplies at the Alliance stores. The latter were to be strengthened by having each county Alliance charter a store of $10,000 capital, half paid in by the local farmers and the other half by the central exchange. As outlined by Macune, the plan actually cost the central exchange nothing in capital, for it acquired the use of the $10,000 capital of each of the county Alliances for half that amount—in effect, using the credit of the local stores. The plan, while certainly strengthening the local operation, had as its principal intent the strengthening of the central exchange; indeed, each county Alliance was to be coerced into subscribing for its proportionate per-capita share—on pain of being excluded from the benefits of the treasury-note plan. The practical effect of establishing a treasury department in the state exchange, one empowered to issue its own currency to Alliance members upon deposit of warehouse receipts of their commodities, was to expand the money supply circulating within the order in a way that strengthened the capitalization of both the central exchange and its local branches.

Macune argued for his treasury-note idea with customary sweep, dismissing the "cash-only" Rochdale inheritance of the Grange as wholly inadequate to the task at hand:

> We have been talking co-operation for twenty years. Now we have made an aggressive movement. It has thrown the whole community into the wildest confusion. . . . It saved us last year from one to five million dollars on our cotton; it saved us forty percent on our plows; thirty percent on our engines and gins; sixty percent on sewing machines; thirty percent on wagons. . . . In spite of all this the question today is: shall we endorse the aggressive movement, or shall we go back home and say to the people, we stirred up the bees in the bee tree and made them make the biggest fuss you ever heard . . . but we declined the fight and have come back home to starve and let our children grow up to be slaves. In a nut-shell then, the question is: will you cease an aggressive effort that promises certain relief, simply because the opposition howl and curse?

Macune's words underscored the fact that he was fighting for his life as exchange manager and national spokesman for the Alliance. But the experience had shaken the farmers. Alliancemen stood loyally by Macune, their "past, present, and future business manager," but they clearly felt they did not have the immediate means to implement the treasury-note plan. In essence, both Macune and the farmers were forced to acknowledge that the treasury-note plan, however ingenious, was beyond the means of the penniless farmers of Texas. The exchange drew up new by-laws relinquishing, at least for the present, the basic struggle against the credit system: "all purchases made from branch exchanges must be for cash, and notice is hereby given that the books of the exchange are closed against any further debit entries." This was a bitter retreat; it reflected the belief that if the cooperative effort were ever to make a second attempt, the exchange had to be placed on a sound footing. The officers betrayed their anxiety by exhorting the membership to pay up the joint notes: "settlements are at once required, and must be made as soon as possible."

Yet neither Macune nor his principal advisers in the Texas leadership appeared to be intimidated by the experience or inclined to shrink from the complexity of managing large-scale farmer cooperatives. At the Alliance's second national convention in Meridian, Mississippi, late in 1888, Macune and another cooperative theorist from Texas, Harry Tracy, argued earnestly with state Alliance leaders that the Rochdale system helped only the thin layer of the agricultural middle class in the South, that it ignored the crushing needs of the great mass of tenant farmers and the hundreds of thousands of landowners caught in the cycle of debt to merchants. Macune insisted at Meridian that the Alliance cooperative had to go beyond the joint-stock Rochdale plans of the past. The central state exchange, he said, "is calculated to benefit the whole class, and not simply those who have surplus money to invest in capital stock; it does not aspire to, and is not calculated to be a business for profit in itself."

> Instead of encouraging a number of independent stores scattered over the country, each in turn to fall a prey to the opposition, whenever they shall think it of sufficient importance to concentrate a few forces against it, this plan provides for a strong central State head, and places sufficient capital stock there to make that the field for concentrating the fight of the opposition, and a bulwark of strength and refuge for the local store efforts.

Alliance delegates could not with consistency argue against an attempt "to benefit the whole class," as the Alliance itself was predicated upon that approach. And they did not. Tracy, Macune's colleague, elaborately spelled out the central state exchange concept and informed North Carolina's L. L. Polk that cooperative stores in every Southern hamlet would accomplish little in the way of "financial liberty" as long as "they buy, ship, and sell independent of each other." In terms of the mass of Alliance farmers, Tracy saw no "utility" in lesser cooperative attempts. Such ideas were inherently radical, of course, though men like Tracy who were allied with Macune still voted Democratic and did not regard themselves as political insurgents.

13

But the searing educational experience of the cooperative struggle of 1888 had by no means been restricted to Charles Macune and the top Alliance leaders. All Texas Alliancemen had gone to this school. It had conveyed some unique lessons about American banking practices. Indeed, to farmer advocates who sought to benefit "the whole class," the dynamics of the cooperative movement had also brought a new perspective on the larger American society. The discovered truth was a simple one, but its political import was radical: the Alliance cooperative stood little chance of working unless fundamental changes were made in the American monetary system. This understanding was the foundation of the Omaha Platform of the People's Party. In August 1888 it materialized in the organization that would carry it to millions of Americans.

At the 1888 convention of the Texas Alliance, the intense intellectual discussion of the future of the exchange—and the cooperative crusade that had generated the climate for such discussions—brought the agrarian movement to a critical ideological threshold. In the very months that Macune and the Alliance leadership labored frantically to anchor in place the cornerstone of the cooperative commonwealth—indeed, in the very weeks the dirt farmers of the Alliance met, marched, and contributed in an effort to save their exchange—Alliance radicals quietly and almost imperceptibly achieved a decisive ideological breakthrough: they conveyed the greenback political heritage to the spreading agrarian movement as the centerpiece of Alliance doctrine.

This seminal development, one that hammered into place the ideological framework of the famed Omaha Platform of the People's Party, was an achievement riddled with irony, for it was accomplished over the objections of the most inventive greenbacker among them—Charles Macune. That it happened at all conclusively verified the ascendency of radicalism within the ranks of the Alliance founders.

In April 1888, as Macune searched in vain for bankers who would honor the collateral of the Alliance Exchange, H. S. P.

Ashby, one of S. O. Daws's original radical organizers, issued a "waking up circular" calling for a statewide meeting "for the purpose of considering what steps, if any, should be taken in the approaching campaign." A leader of the earlier independent political movement which had elected a radical mayor of Fort Worth in the midst of the Great Southwest Strike of 1886, Ashby had become a key participant in a loose coalition of like-minded farmer spokesmen that included Evan Jones, president of the Texas Alliance; J. R. Bennett, editor of the *Southern Mercury,* the order's national journal; and, perhaps not surprisingly, William Lamb. Along with other radicals in the nation in 1888, these men wanted to construct a third-party political institution to replace the defunct Greenback Party. Ashby's meeting, convened in Waco in May 1888 as the "Convention of Farmers, Laborers, and Stock Raisers," promulgated a six-point platform that featured the abolition of the national banking system, the replacement of national bank notes with legal tender treasury notes issued on land security, prohibition of alien land ownership, and "government ownership or control of the means of transportation and communication."

Macune took time out from his Exchange wars to write a lengthy letter to the *Southern Mercury* warning the 250,000 Alliance members in Texas against independent political action. But while Ashby and others held their Waco meeting, Lamb was in Cincinnati assisting in the formal creation of the new national "Union Labor Party." Back in Fort Worth on July 3, 1888, Lamb chaired a statewide "Non-Partisan Convention" which accepted Ashby's Waco platform and added additional planks calling for a federal income tax, free coinage of silver, and the enactment of compulsory arbitration laws. With respect to the railroads, the phrase "government ownership or control" was altered to read "government ownership *and* control." No sooner had Lamb's "non-partisans" finished their work in Fort Worth on July 4 than the "Texas Union Labor Party" held its inaugural convention in the same city on July 5. Amid an amusing—and revealing—overlapping of delegates from the two conventions, the new Union Labor Party adopted—without so much as changing a comma—the platform of the "non-partisans" written two days earlier. The new third party's nominee for the governorship of Texas was no less than Evan Jones, president of the Texas Alliance. As the struggle to save the Exchange reached its climax in the summer of 1888, Jones reluctantly decided that a divisive political campaign would weaken the Alliance at the moment it needed maximum solidarity to preserve the forward thrust of the cooperative movement. Therefore, shortly before the annual convention of the Texas Alliance at Dallas in August 1888, he declined the nomination "of this noble body of men."

But at the convention itself—even as Macune made his lengthy report on the exchange to the attentive delegates—J. M. Perdue chaired an Alliance committee that delivered a radical "report on the industrial depression" that placed greenback monetary theory at the heart of Alliance politics. This document, which concluded with the "Dallas Demands," conveyed to the agrarian movement the radical greenback heritage; the three documents written at Waco, Fort Worth, and Dallas between May and August 1888 provided the entire substance of the St. Louis Platform adopted by the National Alliance the following year. With the exception of one monetary proposal which Macune himself was to provide, the Omaha Platform of the People's Party was in place.

These developments revealed clearly the dynamics that produced the multi-sectional People's Party: the cooperative crusade not only recruited the farmers to the Alliance; opposition to the cooperatives by bankers, wholesalers, and manufacturers generated a climate that was sufficiently radical to permit the acceptance by farmers of the greenback interpretation of the prevailing forms of American finance capitalism. Greenback doctrines thus provided the ideology and the cooperative crusade provided the mass dynamics for the creation of the People's Party. Both reached their peak of intensity in the Texas Alliance in the tumultuous summer of 1888, and the emotional heat from that experience welded radical greenbackism onto the farmers' movement. After 1888, only one step,* a rather sizable

* This step—the final one involving the political education of the mass movement—is the process of politicization described on pp. 91–93 and 125–82. See also Goodwyn, *Democratic Promise*, pp. 649–52, fns. 34 and 39.

one, remained to bring to fruition the creation of a multi-sectional radical party—the conversion of the bulk of the national Alliance membership to the greenback doctrines which had become the central political statement of the agrarian movement. Two men, Charles Macune and William Lamb, working toward opposite purposes, were to provide the tactics that produced the mass conversion.

14

But in 1888 radical politics was not yet the uppermost thought in the minds of the growing army of farmers who met in thousands of suballiances scattered from Florida to Kansas. The attention of the Alliance was focused on the cooperative crusade. The Texans offered the agrarian movement a bold blueprint of large-scale cooperation, and the very promise of this blueprint recruited farmers by the hundreds of thousands. Large-scale cooperatives were dangerous, it seemed, but men like Macune and Tracy, as well as the more radical leadership among the Alliance founders, argued that only broadly gauged cooperatives could acquire the capital strength to combat the array of weapons available to forces of monopoly. Indeed, the more one learned of the options open to commercial opponents of farmer cooperatives, the more imperative centralized exchanges seemed.

One new state Alliance that agreed was Kansas. As the order moved into its greatest period of growth in 1889, the "Kansas Alliance Exchange Company" was chartered as a centralized marketing and purchasing agency. It later began to publish its own newspaper, the *Kansas Farmers Alliance and Industrial Union*.

The National Alliance spread across the Western plains in 1889, through Missouri as well as Kansas, and moved across the Ohio River into Indiana. Alliance organizers invaded Kentucky and established enclaves in Oklahoma Territory and Colorado in the West and Maryland in the East.

As a decade of organizing came to an end, one of the Alliance's most striking achievements involved victory over its announced enemy, a bona fide national "trust." Cotton bagging had traditionally been made from jute, and in 1888 a combine of jute

manufacturers suddenly announced that henceforth jute bagging would be raised from seven cents a yard to eleven, twelve
and, in some regions, fourteen cents a yard. The action laid a
"tribute of some $2,000,000" on the nation's cotton farmers.
Alliance leaders in Georgia, Alabama, Mississippi, South Carolina,
Louisiana, and Florida reacted with vigor. The Alliance convened
a South-wide convention in Birmingham in the spring of 1889
to fashion final plans for a boycott throughout the cotton belt.
The state Alliances agreed on common plans for action and
entered into arrangements with scores of mills across the South
for the manufacture of cotton bagging. Some buyers, particularly
in England, complained about the inferior cotton bagging, a
problem Florida farmers ingeniously solved by arranging to
import cheaper jute bagging from Europe and paying for it
with farm produce consigned to the Florida Alliance Exchange.
In Georgia, a rising agrarian advocate, Tom Watson, helped
fortify Alliancemen for the struggle by delineating the larger
implications: "It is useless to ask Congress to help us, just as it
was folly for our forefathers to ask for relief from the tea tax;
and they revolted . . . so should we." He added, with an eye to
future struggles, "The Standard of Revolt is up. Let us keep it
up and speed it on." Georgia Alliancemen took him at his word.
When North Carolina's Polk appeared at a Georgia Alliance
convention in the middle of the jute war, he encountered an
up-country farmer dressed out in cotton bagging who told him
that 360 Alliancemen in his county had uniform suits of it and
"they are literally the cotton bagging brigade." A double wedding
at an Alliance Exposition in Atlanta found 20,000 Alliancemen
looking on approvingly as "both brides and both grooms were
attired in cotton bagging costumes." Such innovations testified
to the fervor with which state and local Alliances threw themselves into the battle with the jute trust, but its ultimate outcome
depended on less colorful but more demanding organizational
arrangements to substitute cotton bagging for jute at thousands
of local suballiances across half a continent. The farmers of the
Alliance met this test in 1888–89, and the manufacturing
combine, suddenly awash in its own jute, conceded that the
price-rigging scheme had collapsed. In 1890, Southern farmers

were buying fourteen-cent jute bagging for as little as five cents a yard.

This triumph for cooperative purchasing was matched in quality, if not in scope, by innovations in cooperative marketing that emanated from Kansas. At the local level, county alliances formed a variety of marketing and purchasing cooperatives with an élan born of their new sense of collective power. But cooperation in Kansas soon moved beyond the stage of local efforts. In 1889, the Kansas Alliance entered into joint agreements with the Kansas Grange and the Missouri Alliance and, in 1890, with the Nebraska Alliance, to establish the American Livestock Commission Company. The experiment in multi-state marketing of livestock opened in May 1889 with paid-up capital from farmer members totaling $25,000. The effort proved successful from the start. Within six months the commission company had over $40,000 in profits to distribute to its members. The animosity toward the cooperative among commission companies scarcely promised a serene future, however.

These successes were balanced by a crucial failure. In the late summer of 1889, after twenty months of operation, the Texas Exchange, unable to market its joint notes in banking circles and therefore unable to respond to insistent demands from its creditors, went under. The Texas effort, the first to be chartered, was thus the first to fail. The news sent a wave of anxiety through the entire South: was the Alliance dream unattainable? Were the Texas lecturers wrong? As the Alliance grew, so did the burdens of explaining and proving its program of self-help. With increasing frequency the Alliance founders, driven by the difficulties of cooperation, had to explain to farmers that the opposition to their movement derived from the self-interest of gold-standard financiers who administered and profited by the existing national banking system. Greenback doctrines were thus increasingly marshalled to defend the Macunite dream of large-scale farmer cooperatives. By this process radical monetary theory began to be conveyed to the suballiances by growing numbers of Alliance lecturers. The farmers had joined the Alliance cooperative to escape the crop lien in the South and the chattel mortgage in the West, and the failure of cooperatives, particularly the huge model experiment in Texas, spread deep concern through the ranks of the Farmers Alliance. Wherever a cooperative failed for lack of credit, greenbackism surged like a virus through the organizational structure of the agrarian movement. Slowly, the Omaha Platform of the People's Party was germinating.

Charles Macune was not present in Dallas for the death of the exchange, for the ubiquitous Alliance leader now operated out of Washington, where he edited the new Alliance national newspaper. The *National Economist*, underwritten by Texas Alliancemen, became, under Macune's tutelage, easily the best edited journal of agricultural economics in the nation. Circulation soon passed 100,000. As the Alliance organization completed its second year as a national institution, the enigma that was Dr. Charles W. Macune rivaled that of the cooperative crusade itself as a case study in complexity. Macune's belief that cooperation must serve landless tenants and others bound to the crop lien system stamped him as an economic radical, yet he remained firmly traditional on political issues and adamantly opposed to all talk of a third party. Macune was able to summarize all these ideas in a single sentence: "The people we seek to relieve from the oppression of unjust conditions are the largest and most conservative class of citizens in this country."

15

Precisely what the Alliance movement was in the process of becoming could scarcely have been predicted by the farmers themselves from the contradictory events of 1888–89. Only one thing was certain: the Alliance was attempting to construct, within the framework of American capitalism, some variety of cooperative commonwealth. Precisely where that would lead was unclear. More than any other Allianceman, Charles Macune had felt the power of the corporate system arrayed alongside the power of a self-help farmer cooperative. He had gone to the bankers and they had replied in the negative. Though his own farmer associates had said "yes," they could not marshall enough resources to defeat the crop lien system. Macune knew that an exchange of considerably reduced scope could be constructed on a sound basis within the means available to organized farmers. One could avoid the credit problem simply by operating cash stores for affluent farmers.

But while he was an orthodox, even a reactionary social philosopher, and still a political traditionalist, C. W. Macune was obsessed with the need to create a democratic monetary system. The pressure of the multiple experiences that had propelled him to leadership amid the Cleburne schism, to fame as an organizer and national leader during the Southern expansion, to crisis and potential loss of political power over the exchange, and to constant maneuvering against his driving, exasperating, creative left wing—all, taken together, conjoined to carry Macune to a conception of the uses of democratic government that was beyond the reach of orthodox political theorists of the Gilded Age. Out of his need for personal exoneration, out of his ambition, and out of his exposure to the realities in the daily lives of the nation's farmers, Macune in 1889 came to the sub-treasury plan. Politically, his proposal was a theoretical and psychological breakthrough of considerable implication: he proposed to mobilize the monetary authority of the nation and put it to work in behalf of a sector of its poorest citizens through the creation of a system of currency designed to benefit everyone in the "producing classes," including urban workers.

The main outlines of the sub-treasury system gradually unfolded in the pages of the *National Economist* during the summer and fall of 1889. There can be no question that Macune saved his proposal for dramatic use in achieving, finally, a national merger of all the nation's major farm and labor organizations. He now planned that event for St. Louis in December 1889, when a great "confederation of labor organizations," convened by the Alliance, would attempt to achieve a workable coalition of the rural and urban working classes, both North and South.

But the sub-treasury was more than a tactical adjunct to organizational expansion. Macune's concept was the intellectual culmination of the cooperative crusade and directly addressed its most compelling liability—inadequate credit. Through his sub-treasury system, the federal government would underwrite the cooperatives by issuing greenbacks to provide credit for the farmer's crops, creating the basis of a more flexible national currency in the process; the necessary marketing and purchasing facilities would be achieved through government-owned warehouses, or "sub-treasuries," and through federal sub-treasury certificates paid to the farmer for his produce—credit which would remove furnishing merchants, commercial banks, and chattel mortgage companies from American agriculture. The sub-treasury "certificates" would be government-issued greenbacks, "full legal tender for all debts, public and private," in the words of the Alliance platform. As outlined at St. Louis in 1889, the sub-treasury system was a slight but decisive modification of the treasury-note plan Macune had presented to the Texas Alliance the year before. Intellectually, the plan was profoundly innovative. It was to prove far too much so for Gilded Age America.

In its own time, the sub-treasury represented the political equivalent of full-scale greenbackism for farmers. This was the plan's immediate import: it defined the doctrine of fiat money in clear terms of self-interest that had unmistakable appeal to farmers desperately overburdened with debt. As the cooperative crusade made abundantly clear, the appeal extended to both West and South, to Kansas as well as to Georgia. Macune's concept went beyond the generalized greenbackism of radicals such as William Lamb to a specific practical solution that appealed directly to farmers in a context they could grasp. But more than this, the sub-treasury plan directly benefited all of the nation's "producing classes" and the nation's economy itself. For the greenback dollars for the farmers created a workable basis for a new and flexible national currency originating outside the exclusive control of Eastern commercial bankers. Beyond the benefits to the economy as a whole, Macune's system provided broad new options to the United States Treasury in giving private citizens access to reasonable credit. As Macune fully understood, the revolutionary implications of the sub-treasury system went far beyond its immediate value to farmers.

The line of nineteenth-century theorists of an irredeemable currency—one that included such businessmen as Kellogg and

Campbell and extended to such labor partisans as Andrew Cameron—culminated in the farmer advocate, Charles Macune. As Macune argued, the agrarian-greenbackism underlying his sub-treasury system provided organizational cohesion between Southern and Western farmers. As he did not foresee, it also provided political cohesion for a radical third party. The People's Party was to wage a frantic campaign to wrest effective operating control of the American monetary system from the nation's commercial bankers and restore it, "in the name of the whole people," to the United States Treasury. It was a campaign that was never to be waged again.

II
The People's Movement
Encounters the Received Culture

"We are emerging from a period of intense individualism, supreme selfishness, and ungodly greed to a period of co-operative effort."

Kansas Allianceman

"One of the wildest and most fantastic projects ever seriously proposed by sober man."

The New York Times on the sub-treasury system.

"It is socialism. . . ."

Nebraska editor on the founding convention of the People's Party.

4

The National Alliance

Organizing Northern Farmers, Southern Blacks, and Urban Workers

". . . the one most essential thing . . ."

It is one of the enduring ironies of history that established systems of hierarchy rarely find it necessary to rely on sensible defenses as an essential means of maintaining power. Police or other modes of social authority are sometimes necessary, but logic rarely is. Indeed, throughout recorded history, the presence in all human societies of jerry-built modes of thought, behavior, and racial and religious memories have served to help protect traditional elites by strewing complicated psychological and emotional roadblocks in the path of those with unsanctioned but relatively thoughtful innovations. So pervasive have been these habits of thought that established hierarchies have tended to be defended as venerable repositories of good sense when they are, in fact, merely powerful and orderly.

A complementary presumption is that insurgent movements are nonsensical. Indeed, the very thought that an insurgent movement, in its fundamental tenets, may be more than superficial calls into question the usefulness of the established order that resists the movement. Participants in the mainstream of most societies generally find such causal relationships difficult to accept because to do so would challenge their own individual modes of thought and behavior. The existence of a coherent protest movement is, therefore, an awkward fact for any society. For Americans, Populism proved particularly awkward.

These circumstances have created for the student of the agrarian revolt a number of conceptual hazards, securely grounded in the traditions of our history and culture. Primary among them is a generalized presumption about "politics" that proceeds from a deep and largely unconscious complacency about American democracy. This attitude essentially embraces three elements. At bottom is a romantic view about the achievements of the American past. The national experience is seen as both purposeful and generally progressive. The "system," though not without flaws, works. Lingering flaws will ultimately be diminished. Unarguably, this presumption is the conceptual centerpiece of that vast body of writing known as the literature of American history. However true or untrue this presumption may be, it incontestably prevails. Indeed, it is a central presumption of American culture.

Building upon this belief is another intellectual folk custom that contributes to the pervasiveness of American complacency: since the "system" passes the historical test as being generally progressive, a progressive evolution of democratic forms over time is assumed. Complementing this belief (the focus here remains upon the mainstream cultural presumptions of American society) is a generalized intuition concerning the relative ease with which democratic forms have come to exist and are maintained in America. Few central ideas about America have less historical basis of support than this one. Hierarchical structures and hierarchical modes of thought are endemic to the societies mankind has created. Democratic ideas are most likely to be regarded, by the cultural elites of such societies, as insurgent ideas. So regarded, these democratic ideas have, in fact, *become* insurgent ideas—even in societies fancying themselves as "democratic." Such was the fate of Populism in America.

Yet there is another component of contemporary American complacency, one that builds upon the first two. Organic to the mainstream American historical tradition is a mode of analysis that has grown out of a profound underestimation of the difficulties inherent in the formation of mass democratic movements. The point can scarcely be exaggerated. The simple reality is that the hazards facing the organizers of democratic movements such as Populism have been, and remain, awesome. So difficult have mortals found these hazards to be, in fact, that the times in human history when a majority of people in any given society have been able to shake off inherited habits of conduct and find a way to participate in significant democratic politics have been rare. Democratic movements may be relatively coherent and purposeful—and still fail.

All of the foregoing is essential to an understanding of Populism. For aspiring spokesmen of the "plain people" in nineteenth-century America, the obstacles to the recruitment of a mass movement were many, varied, subtle, and, in the aggregate, overwhelming. It is, for example, a fact that most nineteenth-century Americans voted for the Democratic or Republican Party because of sectional loyalties generated by the Civil War. As long as Protestant farmers, North and South, divided over sectionalism, and farmers and workers in the North further divided over religion, the possibility of a mass democratic politics free of decisive hierarchical business influence was a remote one.

But the problems facing agrarian organizers were not restricted to the intense sectional legacy of the war, or to deeply held loyalties and prejudices. Alliance lecturers necessarily rested their case on their analysis of the inequities of the American economic system. While this appeal, particularly in the hands of sophisticated greenback theoreticians, could sometimes take on a certain attractiveness, the patterns of deference that had been instilled in working Americans over generations made autonomous decision-making by the "plain people" something less than probable. For a farmer in Iowa or Illinois to leave the Republican Party in order to become a Populist he had to overcome not only his memories associated with the "Party-that-saved-the-Union," but the enduring and very visible civic presence of that same party in his own time and locale. In the towns and hamlets of the rural North and West in the late nineteenth century, the Fourth of July was a day of Republican celebration. The commander of the local unit of the Grand Army of the Republic could be counted upon not only to rekindle memories of loyal "boys in blue," but to lead the Fourth of July parade itself. And this latter was no small political gesture—for included in the ranks parading the (Republican) flag were the town's aging bankers, ministers, and "plain people" who had fought at Shiloh and Gettysburg; and, year after year, the same civic ritual ensured the loyalties of the younger bankers, ministers, and plain people. The past thus blended into the present as a political statement grounded in patriotism and expressed by one's reaffirmed Republican allegiance. Standing up against one's minister, civic leaders, and economic and cultural models not only tested a person's range of psychological autonomy but his intellectual ability to define what authentic patriotism was. The impediments to political nonconformity were impressive. Sectionalism was not merely a patriotic memory, it traded on received patterns of deference in the present.

Almost precisely the same dynamics applied in the South, where the association of patriotism with the Confederacy was augmented by the sentimental pathos inherent in the Lost Cause. "Confederate Day" rallies were also political statements, a refurbishing of loyalties in which one's captains, colonels, and generals led the civic congregation in reverence to the "party of the fathers." In both North and South, Alliance organizers, as they themselves well understood, were asking a great deal of the people in terms of their own sense of autonomy—even of those sectors of the plain people they most closely identified with as Protestant, white farmers.

For urban Catholics in the cities of the North the cultural hazards to organizational recruitment were even more imposing. What could a Protestant agrarian organizer say to an Irish Catholic factory worker in Boston? What did the cooperative crusade mean to the Germans of Pennsylvania's cities? How could an Alliance lecturer even gain access to them? Indeed, what could he say to them should he somehow succeed?

Finally, the American tradition of white supremacy cast a forbidding shadow over the prospect of uniting black and white tenants, sharecroppers, and smallholders into an enduring political force across the South. The two races shared a single, compelling, economic connection—vast numbers of farmers in both races were shackled to the furnishing merchant. But these Southerners also shared with each other and with Northern plain people inherited patterns of deference. In the South these patterns were fortified each time a farmer approached the counter of the man with credit. The merchant was more than a captain or a colonel; he was the man who controlled the farmer's life. Under such circumstances, what political options did impoverished farmers possess? To what extent could the Alliance cooperative encourage a sense of self in people? How much could fellowship in the Alliance be expected to teach? About economics? About race?

Self-evidently, the dream of William Lamb, Henry Vincent, and other Alliance activists for a "new day for the industrial millions" through the creation of a national farmer-labor coalition was a vision that contradicted the realities of American life. In this sense, the early coalition of the Alliance and the Knights of Labor in Texas in 1886 was quite misleading, because few of the reigning cultural obstacles were present. The Missouri-Pacific track laborer who belonged to the Knights of Labor lodge in Parker County, Texas, was largely indistinguishable from the farmer-member of suballiance No. 87 of the Parker County Farmers Alliance. The railroad worker was, often quite literally, the brother who had come to town to earn wages to help the family survive on the crop lien. In the Western plains—whether in Weatherford, Texas, or Winfield, Kansas—the "laborer" in the Knights and the "farmer" in the Alliance dressed alike, attended the same church, read the same newspaper, nursed similar hopes. If the Alliance, or the Knights, or both of them, could somehow generate enough internal momentum, a "farmer-labor coalition of the plain people" might indeed be an attainable objective in certain homogeneous locales in the West. But the same could scarcely be said for the prospects of recruiting a Slavic steelworker in Pittsburgh, a "patriotic" Republican farmer in Michigan, or a black sharecropper in South Carolina. Throughout human history the creation of a new political culture has always involved far more than the propagation of a platform or the existence of "hard times," and the task facing Alliance organizers in nineteenth century America was truly a momentous one.

The manner in which the people of the Alliance, in 1887–92, gained enough self-confidence to shoulder this burden is an interesting story, in some ways a saga of democratic striving.

But it is also necessarily a story that cannot now be completely rescued from the obscurities of historical causation itself. The broad tactical outlines are clear—how provincial organizers developed the means to carry their cooperative message to the Canadian border and to both coasts, how they attempted to build political bridges to labor leaders in the North and to black spokesmen in the South, how the specific ideology of their platform came into being, and finally, how, in the face of sectionalism, they managed to construct a multi-sectional political party. And there is evidence which measures how the Alliance combated inherited patterns of deference among its members, as well as the process through which the supportive culture of the movement helped encourage the mass aspirations that made the Populist effort such a unique moment in American history. But it is impossible at this late date to get into the mind of a nineteenth-century Alabama Populist, be he black or white, or to know why some workers in Chicago found a way to join the vibrant third party effort there and why others did not. While self-generated or movement-generated political consciousness can be, and often is, intimately related to specifics of social and economic class, it is also a product of elaborate and contradictory cultural patterns. We know probably as much as we need to know about who the Populists were, but we almost certainly will never know remotely enough about the specific cultural tensions generated in the consciousness of millions of individual Americans by the appeal and the challenge of the Populist moment. Available are many insights into the human and organizational dynamics of Populism where these dynamics were strongest, as are a number of other insights about the same components where they were weakest. But we can only be much less precise about that broad spectrum of political consciousness where the two tendencies come together—where "the movement" affected people, but did not decisively affect them. We can, however, be certain of one influence: everywhere, the power of the received culture was present.

Yet even with these necessary qualifications, the Populist moment offers a fascinating matrix of aspiration, indignation, despair, and—always—earnest striving. Their dreams were sufficiently grand in scope that, in failing, they left a story of intense democratic effort.

2

The Alliance assault on the rural bastions of the Northern Republican Party is a study in persistence. The Kansas movement, with its articulate nucleus of veteran greenbackers, effectively augmented by the organizing sophistication of the Vincent brothers, constitutes an untypical case of rapid consolidation. Elsewhere the ascendancy of the Alliance was more gradual, and in some places it did not come at all. In all instances, sectionalism constituted a chronic problem that never went away.

Historians long thought that American Populism was largely a product of the efforts of Western farmers generally and of a specific organization popularly known as the Northwestern Farmers Alliance. But we now know this was not the case. The Northwestern Alliance, a small caucus of aspiring spokesmen formed by a Chicago magazine editor in 1880, limped through the decade without ever formulating either an internal program or an external purpose. It consequently had great difficulty either attracting or keeping members. As late as the spring of 1889 it had a constantly shifting membership of less than 10,000 in its entire claimed jurisdiction. As an institution, the Northwestern group contributed almost nothing to the Populist movement; indeed, it opposed the formation of the People's Party.

Prior to the national ascendancy of the Alliance, substantive organizational activity in the Great Plains was restricted to two local efforts in Illinois and Dakota Territory. They shared a component with the Alliance: both were cooperative movements.

The Illinois example was the result of the organizing efforts of the "Farmers Mutual Benefit Association," which grew slowly but steadily through a half-dozen counties in the Illinois wheat belt as farmers attempted to fashion new methods of dealing with grain commission dealers, railroads, and farm implement wholesalers. By 1887, the FMBA had 2000 members who met in subordinate lodges. These groups elected representatives to a county lodge which met quarterly and exercised the kind of direct coordinating guidance over the locals required for co-operative buying and selling. The FMBA was structurally sound. In 1888 came the founding of a newspaper and a proliferation of circulars, instructions, and cooperative planning; growth became rapid. Having achieved a membership of 40,000, the FMBA met with leaders of the Alliance at Meridian, Mississippi, late in the year and worked out plans for immediate cooperation and eventual amalgamation.

Another clear indication of agrarian energy in the Western plains emanated from Dakota Territory. A farmer named Henry Loucks had begun mobilizing wheat growers into rural clubs in 1884 to protest the monopolistic practices of grain elevator companies and the high freight rates of railroads. In the same months that the early Texas Alliance began experimenting in bulking cotton as part of its cooperative marketing program, Loucks led his Dakota farmers into a series of tentative efforts toward cooperation. Some thirty-five cooperative warehouses were erected in the Territory. These attempts had little success, but Dakota farmers learned a great deal, and by 1888 they had formed a territory-wide cooperative exchange. Essentially a joint stock company which employed Rochdale "cash-only" methods, the Dakota cooperative immediately encountered resistance from implement dealers. Such roadblocks had a radicalizing effect on the Dakota leadership, particularly on Loucks himself, and the problems of cooperation heightened Loucks's interest in the procedures of Macune's National Farmers Alliance. The two orders moved into a close relationship in 1888.

The most important contribution that the Dakotans made to the gathering agrarian movement was a unique crop insurance plan. Administered under the leadership of Loucks' close associate, Alonzo Wardall, the cooperative plan not only saved farmers something above 200 per cent over previous rates, but also proved to be an effective organizing tool. The "Alliance Hail Association" met with determined opposition from insurance companies after it announced plans to offer protection for 21 to 25 cents an acre to farmers who were used to paying anywhere from 50 to 75 cents an acre. The Association insured 2000 farmers and 150,000 acres in 1887 and, as the good news spread, 8000 farmers and 600,000 acres in 1888. Wardall shaped the Dakota insurance plan into a multi-state program under the auspices of the National Alliance and, for good measure, formally joined the order in Kansas. He and Loucks readied plans to have their entire membership follow suit in 1889 by a formal merger.

Flushed with success, Loucks and Wardall then personally exported their cooperative insurance plan to Minnesota, resurrecting in the process a moribund organization known as the Minnesota Farmers Alliance. The feat constitutes one of the more interesting vignettes of the agrarian revolt, for they achieved it through the unlikely organizational auspices of a writer-lecturer, social critic, and part-time politician—Ignatius Donnelly.

Though the small and loosely formed Minnesota Alliance dated from 1880, it had always been more or less a plaything of politicians, and the twists and turns in its political maneuvers had scarcely augmented the legislative influence of its membership, which was less than robust in any case. By 1888 the Minnesota Alliance, after eight years of membership in the Northwestern Alliance, had but eighty active chapters, something less than 2000 members, and no internal program of any kind. It was mostly a forum that politicians visited in season. In the spring of 1889, the order, casting about for help, asked Donnelly to become its official state lecturer. Having just completed a sensational anticlerical, anticapitalist novel entitled *Caesar's Column*, and doubtless seeking a new outlet for his energies, Donnelly accepted the offer. Though he had played little role in bringing the Loucks-Wardall insurance plan to Minnesota, he formed a workable partnership with the newcomers, and the triumvirate of two cooperative advocates from Dakota Territory and an aging writer-politician gradually generated the makings of a genuine agrarian movement in Minnesota. Donnelly's deputy state lecturer wrote from the field excitedly that the Alliance was "catching on" because of the appeal of the Loucks-Wardall insurance program. Donnelly immediately began a promotional correspondence with his many personal acquaintances in the state and soon had fashioned several platoons of agricultural insurance salesmen to double as Alliance organizers.

In common with many dissident politicians, Donnelly believed that reform politics was grounded in the art of stump oratory, and his dalliance with organization-building could easily have ended disastrously. But Loucks and Wardall supervised the campaign, Loucks taking temporary leave from his Dakota duties to move in as manager of the Minnesota Alliance business office while Wardall directed the insurance program. In 1889, through Donnelly's ability to recruit agents and Loucks's and Wardall's experience as agrarian organizers, the Minnesota Alliance became a functioning, though embryonic, cooperative. It therefore acquired new members and, for the first time, the beginnings of organizational cohesion. In the last six months of 1889 the Minnesota Alliance acquired a somewhat uncertain lecturing apparatus, a functioning state office, an official newspaper (*The Great West*), a bank balance, and nearly 15,000 members. But Minnesota farmers did not undertake to form a purchasing or marketing cooperative, at either the state or local level.

While the evidence from Illinois, Dakota, and Minnesota seemed to offer at least some promise for Midwestern farmers, the evidence from Kansas was even more instructive. The South might have seemed responsive to the Alliance message, but Kansas set a new standard in consolidation. The cooperative movement surged across the state. In nine months between the spring of 1889 and the winter of 1889–90 more than 75,000 Kansans joined the Farmers Alliance as the order created local, county, and statewide cooperatives. The new movement quickly drew the support of a network of radical newspaper editors whom Henry Vincent had earlier brought together in behalf of the Union Labor party. John Rogers's *Kansas Commoner,* Stephen McLallin's *Advocate,* and, after a pause reflecting years of Republican orthodoxy, even William Peffer's *Kansas Farmer* all joined the *Nonconformist* in defending the cooperative crusade and exhorting farmers to stand firm in trade. By the end of 1889 the cooperative movement had reached Jerry Simpson's part of the state, and the gloom that had punctuated his earlier appraisals began to disappear. When he spoke now as lecturer of the Barber County Farmers Alliance, new faces appeared in the crowd—farmers who had joined the cooperative movement and for the first time were hearing new, radical political interpretations expressed in a setting congenial to their acceptance.

3

While the Alliance probed ever deeper into the Northern plains in 1889, Charles Macune, in Washington, developed his *National Economist* into the national journal of the reform movement and readied some grandiose plans for a national coalition. Macune hoped to bring under a single institutional umbrella all the major working class institutions in the nation. To this end, he worked out arrangements with the FMBA, the Northwestern Alliance, and, most importantly, the Knights of Labor to convene a "confederation of industrial organizations" in St. Louis in December 1889. Quietly, he put the finishing touches on the *piece dé resistance,* his sweeping proposal for a democratic monetary system which he had decided to call the "Sub-Treasury System."

In his months in Washington, Macune had developed a personal relationship with Terence Powderly, the Grand Master Workman of the Knights of Labor. A quiet, cautious man, Powderly was in no sense Macune's intellectual equal and he soon fell almost wholly under the ideological sway of the agrarian leader. For his part, Macune had been deeply affected by the Exchange struggle of 1888 and by the tenacity of banker opposition to his dreams for large-scale credit cooperatives. While he was as touchy as ever about the obvious third party ambitions of Texas radicals such as William Lamb, Evan Jones, "Stump" Ashby, and J. M. Perdue, he was not intimidated by their greenback interpretation of the American banking system. Indeed, purely in terms of monetary theory (as distinct from third party political tactics) Macune was the most committed and inventive greenbacker in the entire movement.

In strictly intellectual terms, therefore, Macune's outlook permitted him to work quite easily with Alliance radicals in fashioning the national Alliance political statement, given to the country in 1889 as "The St. Louis Platform." In the course of complex organizational arrangements worked out at the national Alliance meeting in Meridian the year before, the order had completed the preliminary institutional requirements necessary to absorb the Arkansas Agricultural Wheel. Until that step could be formally ratified by the Wheel membership in 1889, Macune constructed a three-tiered national institution. The National Alliance was renamed the Farmers and Laborers Union of America in a move to pave the way for amalgamation with the Knights of Labor. Evan Jones, the Texas radical, became president. As the Arkansas Wheel moved into confederation, its president, Issac McCracken, maintained equal status with Jones. But over both organizations was a new parent institution, headed by Macune. To that parent Macune affixed the old name: The National Farmers Alliance and Cooperative Union.

Texans then dominated the national agrarian movement, and an intense radical pressure from the Texas Alliance leadership was exerted upon Macune. He deflected it by agreeing with the radicals on the specific substance of the new "Alliance Demands" at St. Louis. These planks incorporated all of the main features of the three radical documents written under the aegis of William Lamb, "Stump" Ashby, and J. M. Perdue during the height of the Exchange war in Texas the preceding year. The Omaha Platform of the People's Party thus came into being without controversy.

4

At St. Louis the Southern Alliances, possessing as they did a zealous assortment of cooperators scattered through various state organizations, acquiesced in these rather profound ideological refinements, which had the effect of committing the entire national movement to the monetary principles of greenbackism. Macune also had no difficulty in inducing Powderly of the Knights of Labor to subscribe to the same political document. The rising Kansas movement and the Dakota group under Henry Loucks also joined in easy harmony. The seven-plank document opened with a full restatement of the greenback heritage that testified to the "educational" impact of the cooperative movement on the organized farmers.

> We demand the abolition of national banks and the substitution of legal tender treasury notes in lieu of national bank notes, issued in sufficient volume to do the business of the country on a cash system; regulating the amount needed on a per

capita basis as the business interest of the country expand:
and that all money issued by the Government shall be legal
tender in payment of all debts, both public and private.

The other six planks called for government ownership of the
means of communication and transportation, prohibition of
alien land ownership, free and unlimited coinage of silver,
equitable taxation between classes, a fractional paper currency,
and government economy. Excepting only the formal addition
of the sub-treasury plan itself, the national agrarian movement
had achieved, on the leadership level at least, the substance of
the Populist vision.

But for Charles Macune, the most important ideological event
at St. Louis was not the platform but the unveiling of his
monetary system. Macune was thoroughly awake to the fact that
his proposal constituted a candid comment on past agrarian
policy, including his own. He gently pointed out the limitations
of the cooperative crusade: "we must realize that there is a limit
to the power than can be enforced by these methods." Farmers
had three choices, he asserted. They could put their hopes on
"more efficient farming"—the orthodox view offered by non-
farmers, especially those of the metropolitan press. Or they
could concentrate their total energies on cooperation—the alternative then being attempted on a massive scale by the Alliance.
Or, finally, they could adopt a new approach: "by organization,
a united effort can be brought to bear upon the authorities that
will secure such changes in the regulations that govern the
relations between different classes of citizens." In arguing for
changed relations between "different classes," Macune suggested
a conscious raising of the stakes above those being gambled in
the cooperative movement. Macune's plan called for federal
warehouses to be erected in every county in the nation that
annually yielded over $500,000 worth of agricultural produce.
In these "sub-treasuries," farmers could store their crops to
await higher prices before selling. They were to be permitted to
borrow up to 80 per cent of the local market price upon storage,
and could sell their sub-treasury "certificates of deposit" at the
prevailing market price at any time of year. Farmers were to
pay interest at the rate of 2 per cent per annum, plus small

charges for grading, storage, and insurance. Wheat, corn, oats, barley, rye, rice, tobacco, cotton, wool, and sugar were included under the marketing program.

The plan carried far-reaching ramifications for the farmer, the nation's monetary system, the government, and the citizenry as a whole. It shattered the existing system of agricultural credit in the South, and returned to the crop-mortgaged farmer some direct control over the sale of his produce, permitting him to avoid the rock-bottom prices prevailing at harvest on all that he sold. It also, for the first time, gave him flexibility in the selling of his certificates. In effect, Macune had replaced the high-interest crop-mortgage of the furnishing merchant with a plan that mortgaged the crop to the federal government at low interest. It thus provided the farmer with the means to escape, at long last, the clutches of the advancing man and recover a measure of control over his own life. For the farmers of the South, both black and white, the sub-treasury plan was revolutionary.

Its manner of presentation at St. Louis, however, was the first sign that, as a national spokesman, Charles Macune possessed serious provincial limitations. Technically, the sub-treasury plan had been placed before the Alliance delegates at St. Louis in the form of a report from a "Committee on the Monetary System" composed entirely of Southerners. As soon as the Kansans heard about it—in public session—they quickly pointed out that it was their farms, not their crops, that were mortgaged. The provision for low-interest loans on farmlands was easily incorporated into the system. Once their cooperatives began encountering strong banker and merchant opposition—soon after the St. Louis meeting—the Kansas leadership quickly became quite taken with the potential of the sub-treasury.

As amended, the sub-treasury appealed to a multi-sectional coalition of farmers, but it clearly faced stout opposition from other quarters. It was soon estimated by opponents that the plan, if implemented, would inject several billion dollars in sub-treasury greenbacks into the nation's economy at harvest time. In the America of the Gilded Age, neither politicians nor businessmen had yet become used to thinking on such a scale.

After all, only $400 million or so in greenbacks had been printed in the course of the entire Civil War. Also, the sub-treasury seemed to be "class legislation," and thus could be opposed as unconstitutional. Such an argument conveniently overlooked an array of government subsidies to business, from whiskey warehouses to transcontinental railroads, but it was a viewpoint that was to be expressed repeatedly as the sub-treasury came to be discussed nationally.

But the most explosive implication of Macune's monetary theory was also an ironic one: he had fatally undermined his own traditional "nonpartisan" position by introducing the flexible currency system. How could it become law without agitation, followed by legislation? How could it be passed without overt Alliance political support? And if neither major party supported the sub-treasury, what could the creative young spokesman say to third party advocates then? The sub-treasury represented a clear admission that Macune's dream of a farmer-led nationwide cooperative commonwealth had failed—it had failed on its own terms both as an unaided instrument of agrarian self-help and as a purely economic solution to the farmer's woes. The farmers simply lacked the access to low-cost credit necessary to make large-scale cooperatives workable. And large-scale cooperatives were the only kind that could address the needs of the sharecropper, the rent-tenant, the marginal smallholder, and all other landowners unfortunate enough to be trapped by the crop lien system. Similarly, only a large-scale cooperative could save Western farmers from either exploitive interest rates or monopolistic marketing practices at both the manufacturing and wholesale levels.

But since the central state exchange was simply beyond the means of farmers, the sub-treasury had become absolutely necessary. Across the South and West in 1889, neither the leaders nor the members of the Farmers Alliance yet perceived this unpleasant truth. But the Texans had been forced to learn it, and Macune knew it best of all. Underlying the negotiations at St. Louis was a new and desperate reality that had surfaced in 1889: the "mother alliance" in Texas had suffered a trauma in the aftermath of the collapse of the Texas Exchange. The vast agrarian army of 250,000 that had rallied on "the day to save the exchange" in June of 1888 had lost many soldiers in the succeeding months. Though those who persevered in their loyalty to the Alliance were undoubtedly wiser in the ways of wholesale houses, bankers, and party politicians, the incontestable fact was that only 150,000 did so persevere through 1889. One hundred thousand other Texas farmers had tested the heralded "big store of the Alliance," had found it wanting, and had, temporarily at least, crept back into their silent world of peonage.

Macune's plan built upon this bitter cooperative experience and represented a reopening of the exchange conflict on new and stronger grounds. It was on the sub-treasury, not on third party agitation, that the Alliance needed to put its emphasis. Macune wrote in the *Economist* that the Alliance "scattered too much and tried to cover too much ground" and needed to concentrate upon the "one most essential thing and force it through as an entering wedge to secure our rights." The government's capital would put a firm foundation underneath the cooperative crusade—and nothing else could.

His emphasis on the sub-treasury was an acknowledgment of how desperate the farmer's struggle against the financial system had become. On the eve of the Alliance's growth to national status, Macune dedicated himself to persuading the order that the sub-treasury was imperative—he saw it to be the only way the idea of agrarian self-help could be preserved and political adventures avoided. It would take time, of course; the farmer-leaders of functioning cooperatives would scarcely want to hear the topic framed in such a forbidding context of desperate options. But as the cooperatives lost their battle with the banking system, the sub-treasury philosophy promised to become more comprehensible. Indeed, to one such as Macune, who had no faith in a radical third party, his plan was the last real hope for economic justice for the great mass of American farmers. They merely needed to understand it.

He turned to the task with considerable energy. In 1890–91 the *National Economist* became a schoolroom of currency theory. The details of the plan and the relative importance of each of its features were outlined again and again in the months during which the National Farmers Alliance transformed itself into the People's Party.

Important as these developments were to the nationalization of the movement culture, Macune and other Alliance strategists also encountered disappointment at St. Louis. The introduction of the sub-treasury system did not have the galvanizing impact that Macune had hoped, either on the Illinois FMBA or on the small leadership caucus known as the Northwestern Alliance. Alliance strategists wanted merger, for the order urgently needed lecturers in the North and West who had unimpeachable sectional credentials. Though the Texans had organized the South and had brought the same cooperative formula to Kansas, Texas lecturers had not conducted the swift organizing sweep of the Alliance through Kansas; Kansans had. Even so, the Kansans had had to overcome local criticism for being part of a movement led by "ex-rebels." Sectional emotions thus consti-tuted a real organizational barrier, one that the St. Louis meeting was designed to solve.

What the Alliance wanted, and did not get, was permission to introduce its cooperative structure and methods of recruitment into all of the states of the Northern plains left organizationally untended by the Northwestern Alliance. Ironically, what the movement did get was another name change, one that would carry it through the Populist era. In the course of the fruitless negotiations at St. Louis, the Northwestern group proposed that the new institution be called "The National Farmers Alliance and Industrial Union." This was readily agreed to. But the Northwesterners backed away from merger, new name and all. The evidence seems conclusive that they were simply intimidated by the sheer size of the Alliance, by its cooperative subsidiary known as "The Association of State Business Agents," and by their own lack of familiarity with cooperation. Rather than disappear into that maw, they kept their moribund organization and their leadership titles, and went home from St. Louis to historical obscurity. The Alliance itself had to organize Northern farmers. The Kansans and Dakotans would have to lead the way.

5

For the reform movement, an important moment had arrived. It was apparent that two concepts jointly controlled immediate Alliance policy. One, naturally enough, was the cooperative crusade itself. In a number of forms—purchasing, marketing, and even manufacturing cooperatives—the idea of cooperation consolidated its grip on farmers throughout the South and West. Indeed, the days of highest hopes and most ambitious planning were still ahead. From Florida to North Dakota, the cooperative crusade moved into its most intensive period of experimentation. It became, quite simply, the consuming life of the Alliance. In what were now literally thousands of suballiances, farmers learned from their business agents and their county trade committees the nature of the opposition of "the town clique" to their cooperatives. At earnest lectures on the monetary system, they learned how the gold standard directly affected their lives. At other lectures they learned about the sub-treasury system and discussed it with their brethren. They became variously thoughtful, indignant, hopeful, and ready for action. In such ways—and unseen by the rest of the nation—the movement culture of Populism came to rural America.

The second concept, one that fit in neatly with the growing political impulse of the cooperative crusade, was the idea of a national institution of the "producing classes." This, too, had great potential in terms of policy. The goal sprang easily from eighteenth- and nineteenth-century egalitarian thought and drew strength from both individualist and socialist traditions. Though it had not gained the status of doctrinal authority in any significant American institution, the concept of a radical farmer-labor coalition embodied visions of attainable human societies that had transforming impact on those who fell under its sway. It does not require an excessive leap of the imagination to conclude that Alliance national presidents literally meant what they said when they spoke of "centralized capital, allied to irresponsible corporate power" and pronounced it a "menace to individual rights and popular government." State Alliance presidents described causes and effects in ways that cast the American idea of progress in new and ominous terms. "While railroad corporations are penetrating almost every locality with their iron rails, they are binding the people in iron chains." Ostensibly, these corporations were "creatures of law," but in fact they were "unrestrained save by their own sense of shame," and shame was "a virtue rarely if ever found among the attributes of monopolistic corporations." From where these men viewed the nation's economic system, the American population *did* consist of "the masses" and "the classes." The latter were already in harmony with themselves, manipulating both of the old parties and constantly "downing the people." They were, in short, "organized capital."

These two short-run concepts, the cooperative crusade and its accompanying political sensibility, led inevitably to a new democratic imperative: to oppose "organized capital," the Alliance itself would have to organize the masses. The burden fell on the Alliance alone since the Knights of Labor, shattered by Jay Gould and unable to develop any effective new organizing methods under Powderly, had lost its momentum, and the tiny Northwestern farm group had never acquired any.

6

The national organizing campaign of the Farmers Alliance began in the spring of 1890. Alliance lecturers moved across the West to California and up the Pacific coast to Oregon and Washington. Whether with confidence or a controlled sense of trepidation, they moved through West Virginia and across the Alleghenies into Pennsylvania, Delaware, New Jersey, and New York. In mingled Southern and Midwestern accents, they told the veterans of the Grand Army of the Potomac about their new vision of cooperative agriculture in the modern state, and about the sub-treasury plan. The anti-monopoly oratory of the Alliance movement was sent across the Ohio River into Indiana, Illinois, and Ohio and beyond to Michigan and Wisconsin. In the mining and cattle regions of the mountain West—Montana, Idaho, Wyoming, Colorado, New Mexico, and Arizona—farmers in scattered valleys heard traveling lecturers and christened their first local Alliances. "We are trampling sectionalism under our feet," L. L. Polk told Michigan Alliancemen. They gave him "cheers for the speaker and the Alliance." The old radical dream was happening, they told themselves. The producing classes of the North and South were no longer going to vote against each other and keep monopoly in power.

They were heady days, indeed—the organizing months of 1890–91. The National Alliance itself would be the great coalition of the masses. The lecturing system expanded, drawing willing believers who set out to convey the Alliance message to the millions who needed it. Confidently engaging in well-attended public debates in town halls and at outdoor rallies across the nation, they even acquired some nicknames in the process. In the course of demolishing an urbane Republican dandy called "Prince Hal" Hallowell, one Kansas Alliance orator, somewhat less regally attired, conceded that "princes wear silk socks" and confessed that he himself had none. From that day forward he was "Sockless Jerry" Simpson—a sobriquet pinned on him by a hostile Kansas press, but one that carried him to the United States Congress. In the first of a well-advertised series of debates in Kentucky, a rangy Southern apostle mobilized the more radical speeches of Thomas Jefferson, stirred in selected social gospels in behalf of the new faith, and overwhelmed a staid city Democrat. The regional press dubbed the intruder the "Cyclone from Texas" and the nation learned that it would have to cope with a political evangelist known as "Cyclone" Davis. Alliance national headquarters sent him travel money, and "Cyclone" was soon bringing the New Testament, a jacobinized Thomas Jefferson, and the sub-treasury plan to farmers from Colorado to Oregon.

As the lecturing system acquired some stars, Macune pressed state alliances to establish official newspapers as a prelude to something more fundamental in the way of truth-bringing—an alliance of country newspaper editors. Macune's newspaper alliance met an obvious need, and the structure of the "National Reform Press Association" was readied for submission to the Alliance membership at the 1890 meeting. The idea was destined to take hold: the press association, uniting over 1000 newspapers across the nation, in due course became the propaganda arm of the People's Party. The agrarian leaders moved to correct other structural weaknesses uncovered by the scope of their organizing effort. They would need an "Alliance Lecture Bureau" to coordinate what had become a national circuit and volumes of literature to explain and justify the cause. The National Economist Publishing Company in Washington began to pour forth a torrent of pamphlets, broadsides, and books on "the financial question," on "concentrated capital," and on the sub-treasury system. The movement had many forums. Polk's *Progressive Farmer* went to 12,000 farmsteads in North Carolina and the rest of the Southeast, the *Southern Mercury* to 30,000 readers throughout the South, and the *National Economist* to over 100,000 throughout the nation. A dozen journals in Kansas, led by the official Alliance journal, *The Advocate*, and including *The American Nonconformist*, the *Kansas Commoner*, and the *Kansas Farmer*, spread the good news, the indignant and sometimes angry news, that had become the message of the Farmers Alliance. The *Advocate*'s circulation soared toward an eventual peak of 80,000. The teachings of the Alliance, fully grounded in the greenback heritage, began to reach millions of Americans.

7

But through the spring of 1890 most of the new listeners were white, rural, Protestant, and, despite the promising growth in Missouri, Kansas, Colorado, and the Dakotas, most were still Southern. To cope with these assorted problems, the National Alliance stretched its resources to the limit. The most effective FMBA lecturer was recruited and put to work throughout the Great Lakes region. Prominent Knights leaders, including the order's national lecturer, Ralph Beaumont, a socialist, were courted, won, and consulted with. Overtures were made toward Samuel Gomper's newly arrived American Federation of Labor.

But none of the labor institutions had developed (or, in the case of the Knights, any longer possessed) an organizing tool that remotely matched the recruiting and educational power of the large-scale Alliance cooperative. Indeed, it would not be until the concept of the sit-down strike provided the C.I.O. with a similar self-describing value in the 1930's that an American working class institution would be able to achieve an emotional drive comparable to the one that animated the Farmers Alliance during its great organizing sweep of 1887–92. Not until the labor movement developed a tactical solution to the problem of strikebreakers, a solution found only after three generations of experimentation, did effective organization come to a substantial part of the nation's industrial work force. In the 1890's, that tactical breakthrough was still far in the future. As the People's Party was to discover, unorganized workers were hard to wean from inherited political loyalties.

8

One other sector of the "producing classes"—black farmers in the South—had long been a source of Alliance interest. By 1890, definite signs of organizational progress had surfaced. These signs centered on an organization known as the Colored Farmers National Alliance and Cooperative Union. The Colored Alliance was easily one of the most interesting institutions developed in the course of the agrarian revolt. But while much of its evolution was traceable to the actions of black people, its origins were a result of white radicalism.

The order had its source in the political analysis of committed greenback radicals among the Alliance founders in Texas. Though the cooperative struggle to preserve the Texas Exchange eventually brought almost the entire leadership of the founding Alliance to a firm third party position, so that all Texas Alliancemen tended to appear "radical" to the nation, degrees of difference always persisted in their ideological participation in the most radical dream of all—a farmer-labor coalition of the "plain people" that was interracial. A half-dozen years before the formation of the People's Party, the necessary political formulas for constructing such a coalition had become fairly settled doctrine within what might be loosely described as the "Lamb wing" of the Texas Alliance. The politics of sectionalism that had defeated the Greenback Party in the South was to be overcome through the Alliance cooperative; white farmers would join the cooperatives, learn their radical lessons, and then cross the Alliance "bridge" from the Democratic "party of the fathers" to the new "third party of the industrial millions." But the politics of sectionalism quite obviously affected blacks as well. Was not a similar institutional "bridge" needed to convey the black "plain people" from the Republican Party into third party coalition with the white "plain people"?

The massive organizing drive of the Texas Alliance that followed the Cleburne Demands in 1886 generated a self-confidence and enthusiasm among Alliance lecturers that made organization of black farmers a distinct possibility. In the months following Cleburne, as Alliance lecturers enrolled whole farming districts in the "Old South" part of Texas, several black Alliances came into being. One, organized at a farm near Lovelady in Houston County, became the institutional base of the Colored Farmers' National Alliance. Sixteen participants in the meeting were black, and one of them, J. J. Shuffer, was elected president. However, the seventeenth person present, the man on whose farm the meeting took place, was R. M. Humphrey, a white Baptist minister, ex-South Carolinian and ex-Confederate officer. Humphrey was named "general superintendent" of the order and always thereafter served as its chief national spokesman.

It seems clear that Humphrey visualized the organization of black farmers as part of a larger commitment to national radical politics. But it seems equally clear that he had a short-run personal objective as well—a desire to be elected to Congress from his East Texas district, which was heavily populated by blacks. After a measure of organizing success in Texas in 1887, the "Colored Alliance" came into being as a national organization in the spring of 1888—in time for Humphrey's congressional effort. His rather conclusive lack of success in the subsequent campaign seems to have permanently ended his pursuit for higher office. Thereafter, Humphrey's interest in the organization he headed became almost the entire focus of his subsequent life as an Allianceman and Populist. The black organizers who deployed through the South carried the basic cooperative message of the Alliance, and Humphrey later wrote publicly about the state cooperative exchanges established by the order in New Orleans, Mobile, Charleston, Norfolk, and Houston. By 1890 the minister, who had a decided talent for press relations as well as self-promotion, claimed no less than 1,200,000 members in sixteen states.*

Organizers of the Colored Alliance were both black and white, the latter tending to serve in the cooperative exchanges. The black president of the Alliance in Alabama, Frank Davis, was characterized in an Alabama newspaper as "a solvent and successful Negro farmer." Other black Alliancemen who grew to prominence in the movement included E. S. Richardson and J. W. Carter of Georgia, J. S. Jackson and J. F. Washington of Alabama, J. L. Moore of Florida, W. A. Patillo of North Carolina, William Warwick of Virginia, H. J. Spenser of Texas, Joseph H. Powell of Mississippi, and L. D. Laurent of Louisiana. The political world in which these men lived was not simple. To the extent that black Alliancemen made any moves that suggested an eventual political coalition with white agrarians, they threatened a number of existing political arrangements of both races.

The political problem facing Southern blacks was enormously complex, even though it contained few genuinely palatable options. The economic imperatives were quite real—blacks suffered as much from the ravages of the crop lien and the furnishing merchant as white farmers, and more. The black man, too, wanted to find a way to finance his own crop without putting his economic life in the hands of the man with credit. He, too, wanted a more flexible currency, higher commodity prices, an end to high freight rates, and all the rest of the Populist goals. But in an era of transcendent white racism, the curbing of "vicious corporate monopoly" did not carry for black farmers the ring of salvation it had for white agrarians. It was the *whiteness* of corporate monopoly—and the whiteness of those who wanted to trim the power of the monopolists—that worried Negroes. Both sets of white antagonists lived by the values of the American caste system. The rare black farmer with enough capital to stay out of the grip of the crop lien knew quite well that he was just as vulnerable to the whims of Southern justice, just as unprotected against lynch law, as the most downtrodden

---

* Humphrey's claims were quite inflated. A figure of 250,000 seems a fair estimate of the size of the Colored National Alliance. Even this figure is impressive, given the impediments to recruitment then existing. See pp. 122–23.

tenant farmer. In this fundamental sense, economic improvement gave him no guarantee of protection. For black Americans there was no purely "economic" way out. Though every black person in the nation over the age of ten knew this, neither white conservatives who prided themselves on their "tolerance" nor white radicals in the grip of ideological visions of "a new day for the plain people" were aware of their own participation in a caste system that lay at the heart of this crucial question for blacks. The self-serving capacity of different—and competing— groups of whites to dissociate themselves from "the race problem" was a fact that was abundantly clear to black leaders, both North and South. Blacks therefore approached the whole issue of politics from a much more sophisticated—and cautious— perspective than either white conservatives or white insurgents. They, unlike the whites on both sides, did not enjoy the presumption that elections had to do with economic and political power. Before the black man could worry about economic injustice, he had to worry about survival.

Black community leaders needed to find ways to develop safeguards both for themselves and for blacks generally. These leaders knew that no Southern farmer needed to escape from the crop lien more than the black tenant, but they also knew that the economic appeal of the Alliance raised a number of other possibilities, few of them good. First and foremost, black Alliancemen incurred the wrath of Southern white conservatives who exercised the power of governmental authority, including the police authority. The staunchly Bourbon Montgomery *Advertiser*, for example, became offended by signs of fraternization between the black and white Alliances and reminded the members of both that "the white people don't want any more Negro influence in their affairs than they have already had, and they won't have it."

Black Alliance organizers not only had to keep an eye out for white Bourbons; they also had to cope with Negro Republicans. The reform movement threatened the power bases of both groups. In Virginia, North Carolina, South Carolina, Alabama, and Texas, entrenched black Republican leaders systematically undercut the efforts of organizers for the Negro Alliance, and, understandably, Negro Republicans had to ponder the long-range gamble implicit in public identification with the agrarian cause. If these black leaders relinquished their power base within the Republican Party—and it was the sole political base black men had remaining to them in the South—only to discover that the agrarian movement failed politically, they faced the probability of being left with no personal foothold at all in the electoral process. Many black Republicans decided—correctly, as it turned out—that the People's Party was going to lose its battle with the party of white supremacy. Accordingly, they held aloof. The agrarian revolt thus divided both races, whites along economic lines and blacks according to decisions based on cold and necessary calculations of political and physical survival.

Because of the prevailing white racial attitudes, those blacks who led the Alliance in their states frequently functioned covertly. Black lecturers who ranged over the South organizing state and local Alliances did not enter Southern towns behind fluttering flags and brass bands. They attempted to organize slowly and patiently, seeking out the natural leaders in rural black communities and building from there. For this reason, even though the Colored Farmers Alliance was an institution of great range and political significance, its development is shrouded in mystery. Hints of its activities have cropped up in reports of cooperative efforts around Norfolk, Virginia, and in South Carolina, and also in such incidents as the one involving a black newspaper in Vaiden, Mississippi, which was forced to suspend publication during organizing activities in and near the Mississippi Delta. Even more evidence appeared in the form of letters in the *Southern Mercury* and the *National Economist* from newly appointed business agents, occasional press statements by R. M. Humphrey, and the public political activities of the Colored Alliance both at the Ocala convention of 1890 and at the founding meeting of the People's Party in St. Louis in February 1892. All attested to the order's continuing presence across the South.

Nevertheless, white supremacy hung over the organization with a brooding presence that ultimately proved suffocating. The reason was simple: white supremacy prevented black farmers from performing the kinds of collective public acts essential to the creation of an authentic movement culture. Within the Southern caste system, there could be no vast Colored Alliance cooperatives and no public demonstrations of suppport for such cooperatives, no wagon trains stretching for miles, no spectacular summer encampments—nor any of the other public acts of solidarity that help individual self-respect and collective self-confidence take life among impoverished people. In the South of the 1890's, there was almost no public way black Americans could "see themselves" experimenting in democratic forms. Rather, organizing work in the Colored National Alliance proceeded largely in covert ways.

Such constraints are stifling for democratic movements. That the Colored Alliance touched the lives of as many people as it did was a remarkable political and cultural achievement. But it was not remotely enough. By 1890, it was still unclear just how the mass of black farmers across the South were to be recruited away from the party of Lincoln.

9

In general, the arrival of the Alliance as a national force in American life brought into view a paradox at the center of the agrarian revolt. To the extent the challenge of constructing a national farmer-labor coalition confronted Alliance organizers with the intimidating dimensions of the task they faced, it is difficult now to visualize how they avoided chronic despair. Not only did election returns from every major campaign since the Civil War offer continuing proof of the strength of sectional, racial, and religious loyalties among the mass of American voters, but more recent evidence within the reform movement itself seemed just as forbidding. The travail of the Knights of Labor, persisting as it did in the presence of working class hardship, was one sobering political reality; the penetration of sectionalism even into the leadership ranks of farm organizations was another; the racial barrier in the South still another.

Yet no such pessimism afflicted Alliance spokesmen. On the contrary, the lecturers and editors of the movement, led by the new national president elected at St. Louis, L. L. Polk of North Carolina, fairly bristled with energy and optimism. Indeed, an astonishing percentage of the agrarian leadership—Polk and Macune in Washington, hundreds of country editors, and literally tens of thousands of lecturers across the nation—began to talk movingly about something new on the immediate horizon, something vague but portentous, which they characterized variously as "the great contest" and "the coming struggle." They talked about this looming event in different ways because they had different levels of faith in existing American political institutions. Polk and Macune talked of the "sweeping reforms" that were needed and that were coming. Editors like Henry Vincent in Kansas, W. Scott Morgan in Arkansas, Henry Loucks in South Dakota, William Lamb in Texas, and spokesmen like Evan Jones, Jerry Simpson, and a new ally in Georgia named Thomas Watson all talked more frequently of "the coming revolution."

But each group thought in terms of ballots, not bullets. For them, "reform" meant change through the two-party system. "Revolution" meant the overthrow of that system by the creation of a "third party of the industrial millions." Perhaps only the leaders themselves knew the importance of these subtle distinctions; to the rest of the nation all of the Alliance spokesmen sounded alarmingly radical. To many, Macune, with his startling theory of the uses of money, sounded wildest of all.

In the spring of 1890 the National Farmers Alliance and Industrial Union pushed its organizational tentacles to the Canadian border and to both coasts. If Alliancemen and their leaders were asking a bit too much of their farmer organization with respect to the "coming great contest," they did not seem to think so. Grand political dreams are seldom fashioned by cautious men. To many of the agrarian leaders, in any case, there was no time for caution. The farmers of the nation were in trouble, serious and seemingly permanent trouble, and the Alliance had sprung up out of their needs and given voice and meaning to their desperation. Together, in the Alliance, then, they would jointly decide whether to strive for "reform" or for "revolution."

In 1890 the farmers of the Alliance, now well over a million strong, tried both.

5

Reform and Its Shadow
The Core Cultural Struggle

The Alliance calls "the convention of the people"

In the summer and fall of 1890, the National Farmers Alliance appeared to become a mass political institution. In traditional terminology, American farmers became reformers. But the political languages of description traditionally employed in industrial societies are scarcely precise. The Alliance of midsummer 1890 was not yet an insurgent institution because most farmers still thought reform could come through conventional and unreformed agencies of politics—the hierarchically organized and business-dominated major parties. Despite outward appearances, the National Alliance was not yet a democratic political institution; it only seemed so because of the democratic ethos and sense of purpose generated by the cooperative struggle. But this was precisely the limit of the order's mode as of 1890: the Alliance was a democratic institution only in economic terms.

The movement culture, in sum, had yet to "politicize" itself. As long as the great bulk of the Alliance membership thought of politics in conventional terms of major party affiliation, the democratic ethos of the cooperative movement had failed to find a way to express itself politically.

Beneath this reality was an underlying problem of language that has long bedeviled historians and other observers of the process of politics in industrial societies. Should the farmers ever devise a democratic political strategy of their own, one they could regard as straightforward, intelligent, and fully within the American democratic tradition, they would be destined to discover that the larger society regarded them as political pariahs. Indeed, the underlying dynamics of this reasoning process point to the hierarchical nature of American culture itself. In any case, this moment of truth was still ahead for most Alliancemen. With the notable exception of the Kansas Alliance, the only kind of politics most farmers participated in during the campaigns of 1890 was a traditional hierarchical politics.

Indeed, the stormy events of 1890 reflected no consensus among Alliance leaders and no coordination within the emerging national apparatus of state, county and local alliances. The emerging forms of political activity signaled instead an old reality suddenly linked to a new one—the continuing hardship confronting many millions of American farmers and the new collective means they possessed to assert themselves politically. What they wished to assert, however, varied considerably. Collectively, these conflicting procedures offered the first major test of the fragile new movement culture of the agrarian revolt. What the reform movement had not yet become in some places was revealed by what it was already able to achieve in others.

In the Great Plains, the Kansas Alliance offered a case study of a freshly minted mass movement mobilizing itself for democratic politics. On the other hand, farmers in neighboring Nebraska floundered aimlessly and were finally content with halting gestures in hierarchical politics. In a similar fashion, Southern farmers still employed traditional habits of thought in the course of their effort to create a climate of mass democracy. Across the old Confederacy, most members of state and county Alliances unconsciously worked through the only political institution they had ever known: "the party of the fathers." Nevertheless, late in the year the Alliance founders in Texas produced a new tactical weapon conceived by William Lamb and designed to transcend the sectional preoccupations of Southerners. An intense cultural struggle immediately erupted, and within months the politics of Populism surfaced in the Old Confederacy.

In these sundry ways, at different speeds, at varied levels of intensity, and at diverse stages of political consciousness, the farmers brought the People's Party of the United States into being. In so doing, they placed on the nation's political stage the first multi-sectional democratic mass movement since the American Revolution.

2

Most farmers still thought that to change their economic condition it was necessary merely to elect good men to public office. But a smaller number of Alliancemen, convinced that the major parties would never acquiesce in greenback restructuring of the monitary system, looked upon independent political action as the only possible remedy. Though the two views were incompatible, a method was available which gave the appearance of harmonizing them, and it was this expedient which ruled agrarian politics through 1890. For the moment the nation's party system was basically left untouched and the Alliance asserted itself politically by endorsing "Alliance candidates" for office. The tactic, though national in scope, evolved as a series of local decisions that gave the appearance of consistency.

Yet it is perhaps more instructive to view the workings of the agrarian revolt from the perspective of those who did not look upon the politics of 1890 as a unified whole, but perceived instead the deep-seated tensions it really contained. For the significant politics of 1890 represented a tortured groping toward a new political institution of "the plain people." At the beginning of 1890 this view was confined to Alliance radicals, the men who wished to create a new third party in 1892, and certainly by 1896. Though these greenbackers were indeed able to create their party in 1892, their viewpoint, while fervently held, was not a confident one during 1890. Greenbackers had, after all, been third party men for a long time; many had presided over the birth and death of both the Greenback and the Union Labor parties. More than most Americans they knew the difficulty of attracting large numbers of voters to the idea of independent political action or, for that matter, to sustained thought about basic monetary reform itself. Radicals, viewing the political task they confronted after twenty-five years of neo-

Confederate electioneering in the South and the politics of the "bloody shirt" in the North, considered the American voter to be culturally incapable of thinking about politics in terms of ideas that reformers would regard as serious. Too many Americans, including Alliancemen, were, as Evan Jones put it, "still wedded to the 'good old party.' " Though monetary reformers had taken renewed hope from the huge growth of the Farmers Alliance and held a more sanguine view in 1890 than they had at any time in previous years, a number of questions still had to be addressed. Could the educational program of the Alliance overcome the inherited political innocence of the American people? Could the cooperative movement be counted on to spell out the power relationships between bankers and non-bankers that were embedded in the American monetary system? Could a broad enough lecturing apparatus be constructed to support such a burden of political interpretation? Such questions were not easily answered, as the Greenback and Union Labor experiences had proven.

Beyond such ideological and tactical hazards, a nagging structural problem persisted. America is a big country. The immediate task in the West centered on finding some way to initiate a genuine agrarian movement in Iowa, Nebraska, Illinois, and Wisconsin, one that would complement those incubating in Kansas, the Dakotas, and Minnesota.

South Dakota's Henry Loucks appeared before the still largely unorganized Nebraska group early in 1890 to plead for union, but he ran afoul of local leaders who were slow to grasp the essentials of the cooperative movement. At the beginning of 1890, the nub of the problem remained what it had always been in Nebraska—the relative narrowness of the order's organizational base. After almost ten years in the Northwestern group the Nebraska Alliance still had less than 15,000 members in the entire state. The principal difficulty was that the Nebraskans had no program, a fact that did not augur well for holding the relatively few members who had been attracted.

The most telling influence on the Nebraska leadership proved not to be the prodding of Henry Loucks but rather the force of the agrarian movement in Kansas and the Dakotas. This blizzard of activities eventually caught the attention of Nebraska farmers, and the latter, after considerable difficulty, finally caught the attention of their spokesmen. This revolution from below and from without added a certain impromptu style to the stormy politics that shook the South and West in 1890. It created a shadow movement of Populism. A number of Nebraskans dutifully trailed their neighbors in Kansas to the People's Party, but they did so with no understanding of the cooperative crusade of the Alliance or of the greenback doctrines that the cooperative struggle had raised to preeminence in the agrarian movement. In essence, the Nebraskans remained outside the movement culture of Populism. To trace the evolution of the Nebraska shadow movement, it is necessary to begin where the culture of Populism *was* developing—in Kansas.

3

For well over a year the Kansas Alliance had experimented with widespread cooperative programs, and the momentum of the effort had begun to produce a noticeable increase in the political awareness of the state's farmers. The Cowley County cooperative chartered by Henry Vincent and others in 1889 had been followed by the organization of a statewide Kansas State Exchange and a multi-state cooperative livestock marketing plan. By 1890, membership topped 125,000. Alliancemen who had resolved to establish local trade committees had begun to encounter firm merchant opposition. This first stage of cooperation by farmers and its predictable response by merchants also had a predictable educational effect: members of many Alliances became so upset with their merchants that they began organizing their own cooperative stores. A dairy farmer named John Otis declared in January 1890 that the farmers of western Kansas were "burning corn for fuel, while coal miners and their families in another section of our land are famishing for food." Alliance President Ben Clover drew attention to a single law firm in his region of the state that had a contract to foreclose 1800 mortgages! An observant Kansas Republican caught the proliferating signs of restlessness and decided "the air is full of lightning." It was a perceptive observation. The "lightning" was cultural—public flashes of new insights that Kansans were gaining in their suballiances. Examples abounded. On Saturday, the farmers' day in town, the members of the Douglas County Alliance gathered at the courthouse at Lawrence early in 1890 to consider a cooperative crop-withholding program designed to achieve better prices. The farmers elected a committee to pursue the matter. An Alliance county officer then reported gloomily that prices received by local farmers for corn and cattle had fallen below the cost of production. This information produced a series of corroborating stories of personal hardship from the assembled farmers, and a general discussion of prevailing injustices followed. It succeeded only in convincing all present that something had to be done. They resolved "to find out and apply the remedy." Three days later, farmers in Shawnee County, entrapped by the same economic realities, moved beyond resolutions of inquiry to a discussion of political solutions. The unanimous decision of the group, according to one writer, "was that the farmer was being slighted by the legislator, who used him only as a voting tool so that he, the politician, could serve the interests of other classes." In such ways, Kansans created in their suballiances their own environment to think in. In rural forums of democracy, the farmers discussed the difficulties their cooperatives were encountering and moved to "apply the remedy."

The various examples of political lightning observed by Republicans were also fully visible to state Alliance leaders. These activists clearly felt that the movement toward independent political action needed only a small boost. At Topeka, they convened a "State Reform Association" which advertised itself as being in search of a "union platform." The platform duly materialized with the full panoply of Alliance reforms, including the sub-treasury plan. A week later Alliance state president Clover convened a statewide meeting of county Alliance presidents. They undertook an aggressive move that surprised the state's political observers. By all odds the most prestigious Kansan in Washington and perhaps the best known politician in the West was a venerable patriarch named John J. Ingalls, the state's senior United States Senator. The Alliance county presidents resolved that after eighteen years of Ingalls "it is a difficult matter for his constituents to point to a single measure he has ever championed in the interests of the great agricultural and laboring element of Kansas; and we will not support by our votes or influence any candidate for the legislature who favors his reelection to the United States Senate."

Dismayed by this development, the Topeka *Daily Capital* noted that most delegates seemed "strongly anti-Republican." The paper observed, optimistically, that "a large majority of the farmers of Kansas are republicans, and while they are ready to join cooperative organizations for the mutual benefit of producers, they will not consent to being led [by] wild theorists who expect to cure all the ills of financial depression by defeating the republican party." The rebuke did not draw the desired response. An emerging editorial spokesman for the Alliance, Dr. Stephen McLallin of *The Advocate,* announced new political criteria: "We have ends we mean to secure. . . . "To the re-election of John J. Ingalls, for instance, we will never consent. He has been in office eighteen years, but he has got to go."

Precisely how the Alliance intended to elect proper candidates while "no longer dividing on party lines," as the Topeka meeting specified, posed a problem in a statewide organization filled with farmers of Democratic and Union Laborite backgrounds, as well as loyal Republicans. It was decided an "Alliance ticket" should be fielded, but lest such an action be construed as placing "the Alliance in politics" plans were implemented to hold a June convention inviting all "interested citizens." It was the old Daws formula of pre-Cleburne days.

To cope with this unexpected upsurge of democracy, the business community of Kansas countered with a time-honored technique. The "bloody shirt" remained the serviceable, all-weather foundation of Republican electioneering. The brunt of this assult fell not on state Democrats, who were not a threat, but on Alliancemen, who were. As the Topeka *Daily Capital* pointed out, the Alliance was, after all, a Southern-based institution; language usually reserved for Democrats could therefore be applied with equal force to the farm organization. Respectable

Republican farmers, the *Capital* charged, were being duped into doing the work of the Democratic Party at the behest of an organization "officered by rebel brigadiers." The *Weekly Kansas Chief* added that "In Democratic communities, the Alliance is flourishing . . . the most nauseating sight is the course of a lot of Republican papers that are pandering to this organization. . . . Clover is a shyster, a fraud, and an anarchist." The Kansas Alliance thus found what Southern Alliancemen were soon to discover in their own part of the nation—that the agrarian reform program ran headlong into the politics of sectionalism.

In this setting, the much-publicized appearance before the Kansas Alliance convention of the order's national president, L. L. Polk of North Carolina, was watched with anticipation by the state's Republican press. Leaders of the new political movement also watched, but their interest was edged with anxiety. If the reform cause were truly to have a multi-sectional base, it would be crucial that their Southern colleague prove politically acceptable in Kansas.

As it happened, no man in the South was better equipped to cope with this kind of challenge than the fifty-four-year-old North Carolinian. Polk's family heritage and personal political life had instilled in him a driving determination to preach against sectionalism. He had been a Whig as a young man, and had first been elected to the North Carolina legislature as a Unionist in August 1860. That circumstance had profound impact on his life. Opposing secession, he "went with his state" into the Confederacy, only to suffer from repeated discrimination because he was marked as a "Union man." Wounded at Gettysburg, he returned to duty, became an Army candidate for the legislature, and was elected in August 1864. One month following his election, Polk was arrested and charged with "misbehavior before the enemy" and "absence without leave" for having taken his wounded captain to a hospital during a skirmish. Polk was acquitted and serenaded by his regiment before his triumphant return home to legislative service. But the affair was grist for demagogues and it haunted Polk the remainder of his public life.

In the postwar years Polk spoke for agrarian interests in North Carolina, founded the *Progressive Farmer,* and led in the formation of Farmers Clubs and a state agricultural college. When Alliance organizers came to North Carolina in 1887 he watched closely and quietly, and then threw his support to the cooperative movement. Polk became an immediate Southwide spokesman for the Alliance, was elected vice president of the National Alliance in 1887, and became chairman of the national executive committee in 1888. By the time of his election to the Alliance presidency in St. Louis in December 1889 he had made the *Progressive Farmer* the most influential newspaper in his state and had established himself as an outspoken reformer. Thus, by heritage and temperament, as well as through the experiences of his personal life, the Alliance president was ideally suited to cope with the sectional imperatives of national agrarian politics. He roamed the nation from coast to coast in 1890 and habitually culminated his declarations for reform by an attack on the sectionalism that still guided the voting habits of millions of Americans.

Polk's performance in Kansas in the summer of 1890 demonstrates clearly the extent to which the order's leaders were caught up in a momentum which they themselves had helped to generate but which seemed to have acquired a life of its own. To hear the Alliance president, a crowd of 6000 gathered in Winfield on the Fourth of July. It was a new kind of Independence Day celebration, one designated as "Alliance Day." Indeed, the farmers treated townsmen to a new spectacle, one that would grow into a Populist folkway—the Alliance wagon train. The caravan of farm wagons entering Winfield stretched for miles, bearing men, women, children, and provocative political banners.

L. L. Polk, too, embodied the movement. His address focused on the need for economic reform, but his argument was carefully constructed to attack the received traditions of American politics.

> I tell you this afternoon that from New York to the Golden Gate, the farmers have risen up and have inaugurated a movement such as the world has never seen. It is a revolution of thought. A revolution which I pray God may be peaceful and bloodless. ... The farmer of North Carolina, Georgia, Texas, South Carolina is your brother. ... Some people have stirred up sectional feeling and have kept us apart for twenty-

> five years . . . and tried to work upon our passions. The man
> who has waved the bloody shirt. The man who has taught his
> children the poisonous doctrines of hate. . . . They know that
> if we get together and shake hands and look each other in the
> face and feel the touch of kinship, their doom is sealed. I
> stand here today, commissioned by hundreds of thousands of
> Southern farmers, to bid the farmers of Kansas to stand by
> them.

Polk handled the delicate question of "the Alliance in politics"
with the poise and confidence that augmented his personal style
of radicalism:

> Will you tell me who has a better right in America to go into
> politics than the farmers? . . . I will tell you what you are going
> to see. . . . You will see arrayed on the one side the great
> magnates of the country, and Wall Street brokers, and the
> plutocratic power; and on the other you will see the people
> . . . there shall be no Mason and Dixon line on the Alliance
> maps of the future.

To Polk, the "masses" and the "classes" were separated by a
fundamental difference of economic philosophy, one which
measured the selective vision of the wealthy. Sobered by the
arguments advanced in Washington against the sub-treasury
plan, he shared with Kansas farmers his sense of indignation at
the inability of Congressmen to see the conditions of life around
them.

> When I went up to Washington City and showed them statistics
> from all over the country, they said it was overproduction that
> had caused our trouble. . . . If [they] had come out onto the
> streets of Washington on a cold November morning [they]
> would have seen the children picking bits of coal out of the
> ash piles to warm themselves by, and morsels of food out of
> the heaps of garbage to satisfy their hunger. . . . As long as a
> single cry for bread is heard, it is underproduction and
> underconsumption. . . . There is something besides over-pro-
> duction that has caused it. Congress could give us a bill in
> forty-eight hours that would relieve us, but Wall Street says
> nay. . . . I believe that both of the parties are afraid of Wall
> Street. They are not afraid of the people.

Reduced to its essentials, the language of the Alliance was the same, North and South—and it was a language not often heard in mainstream American politics. On the Western plains in 1890, L. L. Polk proved "acceptable." He was invited back for the fall campaign, and he came. The "bloody shirt" was met with economic radicalism and Polk was hailed by Jerry Simpson as "the conductor" of the "through train" of the Alliance that was "going through to Washington." Polk's appearances provided, in the words of W. F. Rightmire, another Kansas leader, "the quasi-endorsement of the National Alliance to the political movement."

But while indignation and sometimes bristling anger was a part of the message of reform, there was another and far more elusive ingredient at work in Kansas in 1890. It has been called a "pentecost of politics," a "religious revival," a "crusade," and it was surely all of those things. But it was also long parades of hundreds of farm wagons and floats decorated with evergreen to symbolize "the living issues" of the Alliance that contrasted with dead tariffs and bloody shirts of the old parties. It offered brass bands and crowds "so large that much of the time it was necessary to have four orators in operation at one time in order for all to hear." It was 2000 bushels of wheat being donated by hard-pressed farmers to help finance their political movement. And it was parades composed simply of the Alliance itself. Some industrious soul counted, or said he counted, 7886 persons and 1500 vehicles in one six-mile-long procession through the city of Wichita. One wonders how the townsfolk of Wichita regarded this vast tide of people. Were they intrigued? attracted? frightened? With parades, speeches, schoolhouse debates, brass bands, Alliance picnics, the politics of Populism took form in Kansas in the months of 1890. If Texans had led the farmers to the Alliance, Kansans led the Alliance to the People's Party.

Yet, in its deepest meaning, Populism was much more than the tactical contributions of Kansans or Texans. It was, first and most centrally, a movement that imparted a sense of self-worth to individuals and provided them with the instruments of self-education about the world they lived in. The movement taught them to believe that they could perform specific political acts of self-determination. The Alliance demands seemed bold to many other Americans who had been intimidated as to their proper status in the society, and the same demands sounded downright presumptuous to the cultural elites engaged in the process of intimidation. But to the men and women of the agrarian movement, encouraged by the sheer drama and power of their massive parades, their huge summer encampments, their far-flung lecturing system, their suballiance rituals, their trade committees and warehouses, their dreams of the new day of the cooperative commonwealth, it was all possible because America was a democratic society and people in a democracy had a right to do whatever they had the ethical courage and self-respect to try to do. Unveiled in Kansas in 1890, then, was the new democratic culture, one created by the cooperative movement of the Alliance.

4

The August convention to nominate the candidates who would carry the Alliance banner statewide was called "the convention of the people." It was a good description. The men chosen had come up out of the ranks of the Alliance—county lecturers, county presidents, some with a long greenback past, others only recently recruited to the Alliance cooperative and to insurgent politics. The gubernatorial candidate was John Willits, the county Alliance leader who had pronounced cooperation "the most important word in the English language." The third district's congressional candidate was the Alliance state president, Ben Clover. The fourth district's congressional candidate was a socialist named John Otis who had earlier written Clover an open letter in which he stated, with appropriate capital letters: "When the American people shall introduce cooperation into the field of PRODUCTION as well as into the field of DISTRIBUTION, and shall organize for 'work' as we organize for 'war'! then shall we behold PROSPERITY. . . ." Otis amplified this theme in the ensuing campaign: "We are emerging from an age of intense individualism, supreme selfishness, and ungodly greed to a period of co-operative effort." Otis provided "perhaps as intense and sober a personality as Kansas Populism counted among its leaders." An abolitionist in his youth, he had attended Williams College and, later, Harvard Law School. During the Civil War Otis organized and commanded Negro troops, but after Reconstruction he turned his back on the party of Lincoln and became a greenbacker. In the western part of the state William Baker brought a different style to Alliance doctrines. A fifty-nine-year-old ex-Republican, he was the most deadly serious and, perhaps, the dullest Populist of them all. Yet he also proved to be one of the third party's most formidable local campaigners.

The pedestrian speeches of Baker and the sustained intensity of Otis were balanced—perhaps more than balanced—by the flamboyant, idiosyncratic, and vividly effective oratory of Jeremiah Simpson. "Sockless Jerry" was an endless source of debate. There was, first of all, the matter of his intelligence. Some years after the third party threat had been safely repelled, William Allen White, an arch-enemy of Kansas Populism, confessed that Simpson had persuaded him to read Carlyle and added, in explanation, that Simpson was better read than he. Be that as it may, there were others who pointed out that self-taught Jerry Simpson was easily one of the most atrocious spellers ever to grace the halls of Congress. In any case, both his friends and his enemies agreed that the mere act of placing Simpson on a rostrum transformed him—but whether into a "good talker" or into "a rabid fiat greenbacker with communistic proclivities" seems to have depended on the observer. He was also, as events were to show, a decidedly pragmatic fellow.

When L. L. Polk returned to the Kansas hustings for the autumn campaign, he found a political tempest that had been gathering in intensity for months. The Republican attacks set a new standard in vituperation. From July through election day the Topeka *Capital* kept up a running "bloody shirt" attack on Polk. He was described as an "ultra-secession Democrat" who had shot down federal prisoners in cold blood at Gettysburg and had practiced barbarous cruelties on Union soldiers while commandant of Salisbury prison—a post, it might be noted in passing, that he had never held. The *Capital* also circulated the story that the "old soldiers of Wichita" were threatening to tar and feather "The Escaped Prison-Hell Keeper."

The Republicans unlimbered other vintage artillery to counter the reform cannonading on "living issues." Simpson was variously described as "unpatriotic" and a "swindler." For the religious-minded, he was "an infidel" and an "atheist." As a politician, he was an "anarchist," and as a human being, he had "simian" characteristics. Vituperation proved to be general. When Ralph Beaumont of the Knights of Labor traveled with Polk through Kansas, they were described as "designing wicked mountebanks," "tramps," "worthless schemers," "enemies of God and man," and "would-be revolutionists." Perhaps worst of all, they had "hellish influence."

On election day, Willits was narrowly defeated for the governorship, but hardly any other Alliance candidate lost. Simpson, Baker, Otis, Clover, and a fifth Alliance candidate, John Davis, all were elected to Congress. The returns carried an especially ominous portent for John J. Ingalls. In that era before the popular election of Senators, his future in the Senate would depend on a Kansas legislature in which 96 of the 125 seats were to be occupied by Alliance-elected candidates.

On the morning after the election it was clear that some sort of earthquake had occurred. By any standard, the Republican Party was a shaken institution. The tremors reached all the way to Washington, where President Harrison was moved to describe the Republican performance in the West as "our election disaster." "If the Alliance can pull one-half of our Republican voters," he said, "our future is not cheerful." It was not necessary, of course, for agrarian insurgents actually to win elections to directly affect Republican fortunes. Every independent vote cast in the West, whether it helped elect radicals or not, weakened the G.O.P. The startling news in states other than Kansas was that the decrease in the Republican vote was enough to send a flood of Democrats to Washington. The totals were sobering. While in 1888 the House of Representatives had had 166 Republicans and 159 Democrats, the new House would contain 88 Republicans and 235 Democrats.

Yet it was precisely this last statistic that sent a wave of anxiety rolling over the newly victorious Kansas Alliancemen. The effect of the revolt in Kansas had been to give convincing meaning to the Republican claim that the agrarian movement was a Democratic plot and that Republican farmers were being duped by the Southern leaders of the Alliance. In the aftermath of victory, the Kansas reformers suddenly found themselves faced with an immediate crisis: they absolutely had to persuade the Alliance in the South to abandon "reform through the Democrats," and create, with the Western Alliances, a national third party. In the West, it was the only defense against the sectional argument that the Alliance was a "front" for the Democratic Party. The convention of the National Alliance, scheduled for Ocala, Florida, in December 1890, thus became of decisive importance to the Westerners.

5

The Kansans soon realized, however, that they had precious few allies in the West to assist them in their earnest internal discussions with their Southern brethren. The movement toward independent political action in Nebraska provided stark evidence of the price the agrarian movement had begun to pay for the Northwestern Alliance's years of drifting. For if Kansas had achieved a revolution, at times flamboyant but politically coherent, Nebraska had engaged in an ad hoc brand of politics comprised of about equal portions of indecisive leadership and organizational anarchy.

It was not that Jay Burrows and his Nebraska associates lacked significant guideposts. As early as 1886 farmers in Custer County, Nebraska, had organized themselves into a local-level cooperative movement, and in 1889, Custer farmers moved boldly into politics in an effort to gain relief. The revolt was centered in the town of Westerville, where each participant in a relatively new cooperative store owned a $10 share and "almost every man in the community belonged to the organization." The movement toward cooperative buying and selling generated outright hostility from Custer County merchants, causing Alliancemen under the leadership of a relatively unknown but articulate farmer named Omar Kem to put a county ticket in the field for the fall elections. When Burrows, the leading agrarian spokesman in the state, belatedly heard details of this revolt from below in the name of the Alliance, he asserted in his newspaper that it was "to say the least, a breach of faith." Awkwardly enough, in November the Custer independent ticket nearly swept the field, all but one of its candidates being elected.

The startling local victory provided one of three influences that were to propel the Nebraskans into independent politics in 1890. The second was the larger cooperative crusade of which the Custer County movement was but a local expression. Kansans had attempted to provide the earliest guidance in 1889, when Henry Vincent spelled out some relevant details of the cooperative movement in Cowley County, Kansas ("4,000 members and a cash business of $300 to $1,400 per day" in the exchange store. Vincent added, "It is tee-totally revolutionizing everything and politicians are, as never before, wholly at sea." But the Nebraska leaders simply did not grasp the connection between cooperatives and politics. Nevertheless, an important additional pressure toward cooperation came from the Kansas and Dakota Alliances. The multi-state Kansas cooperative to market livestock had reached a $2.5 million level of business in 1890, furthering the education of Midwestern farmers, including Nebraskans. Meanwhile, the Dakota Alliance introduced the successful Dakota crop insurance program into the state. Both innovations provided something tangible for Alliance lecturers in Nebraska to talk about—something upon which they could build a movement. Collectively, these outside cooperative influences affixed upon the structureless Nebraska Alliance the appearance of a program and the reality of a promise. And promise, for the time being, was enough. As had previously been the case in Texas, the South, Kansas, and the Dakotas, the farmers in Nebraska now had a reason to join the Alliance. In the spring of 1890, after ten years of sleepy existence, the Nebraska Alliance began to stir.

The third and final influence on the Nebraska leadership came from outside the ranks of the Alliance, from a number of active Republicans who had become disillusioned by railroad domination of the Republican Party in the state. Especially exasperating was the presence of "oil rooms" in Lincoln when the state legislature was in session. Run by lawyers and other lobbyists for railroads, these rooms were for the purpose of "oiling" legislators to vote correctly on pending legislation in which railroads were interested. Apparently, oiling at times extended beyond mere alcoholic lubrication to include matters of finance.

In the spring of 1890, a disenchanted Republican spokesman, Charles Van Wyck, decided that Nebraska needed to cross the same political Rubicon that the Kansas Alliance already had. Three weeks after the county presidents of the Kansas Alliance had signaled the movement of the order into independent political action, Van Wyck decided to show Nebraska leaders how to lead. In a well-publicized address he announced his support for the Alliance, demanded "the abolition of party lines," and otherwise made it clear that the proper business of the Alliance was independent political action.

Omar Kem of Custer County added weight to the pressure. Indeed, Kem's role was crucial, for while ex-Republican politicians might lure the Nebraska Alliance into insurgency, someone first had to recruit the farmers into the Alliance. The Custer cooperative store provided the inspirational model, as the long pent-up grievances of Nebraska farmers at last found a forum for expression. Through April and May, as Kansas moved in measured strides toward independent political action, Jay Burrows was forced to maneuver frantically to keep control of a farmer movement suddenly alive after ten years of passivity.

Finally, Burrows acquiesced to a test of popular sentiment through the circulation of petitions calling for independent political action—though admittedly, the specific goals of such a movement had yet to be fashioned in the state.

Nevertheless, some 5000 signatures were obtained in Custer County, matching an additional 5000 collected across the remainder of the state. Veteran greenbackers and anti-monopolists added their voices, and the resultant din forced Burrows to call an official convention. In July an assemblage of delegates from seventy-nine counties nominated an independent ticket on a platform that emphasized railroad regulation. In a move that Burrows must have found galling, Omar Kem received the congressional nomination from the third district.

During the upsurge from the grass roots that characterized Nebraska farmer politics from November 1889 through July 1890, the Nebraska organization enjoyed all of the growth it was ever to attain in the state. Membership totals peaked somewhere in the vicinity of 35,000 to 45,000. The addition of suballiances fell off sharply after mid-summer and—in the absence of organizational structure—membership quickly began to dwindle away. Local cooperative efforts, undirected and uncoordinated by the state organization, either never actually got started or met some sort of early difficulty and quietly expired even as the political movement gathered momentum with the addition of radicals who had up to then been bystanders. Within the Alliance itself, many new members found they had nothing to do—or nothing anyone showed them how to do— and they shifted their attention to the political movement. In 1890, the Nebraska Alliance thus served primarily as a sort of revolving door to some unspecified kind of insurgent political activity. The decline in Alliance membership continued through 1891.

The ultimate problem in Nebraska was the absence of the kind of statewide cooperative infrastructure that elsewhere provided the agrarian movement with its vehicle of organization, its schoolroom of ideology, and its culture of self-respect. Cooperation, sometimes merely the promise of cooperation, could attract farmers to the Alliance and, under other additional influences, could propel them toward an insurgent political stance; but only the cooperative experience provided the kind of education that imparted to the political movement the form and substance of the greenback heritage. It was banker opposition to large-scale cooperatives that made farmers want to do something about private banking control over the nation's currency. This organic Populist insight into the structure of American society and American politics simply was not present in Nebraska in the years after 1890 because it had not been part of the internal organizational program developed in the North-western Alliance in the years before 1890. Thus, most of the state's farmers were trying to achieve political solutions within a framework of farmer-creditor relationships they had only barely begun to analyze. From beginning to end, the agrarian movement in Nebraska stamped itself as organizationally shallow and ideologically fragile.

In the short run, the absence of clear definition seemed to help the independent political movement, for it induced thousands of voters to support its vaguely defined "anti-monopoly" position in November 1890. The "independents" elected Omar Kem to Congress, joined with Democrats to elect a "fusionist" Congressman, and took consolation in the fact that the state's other Congressman would be a Democrat rather than a traditional regular Republican. The newly elected Democrat—also a respecter of the political uses of imprecision—took "the farmer's side" on the tariff, and was a crowd-pleasing speaker; in the vague climate of 1890 Nebraska politics, the combination proved sufficient to send him to Washington. In that manner the nation's politics acquired a new personality—William Jennings Bryan. The young Congressman was able to work easily with the Nebraska independents, for they all shared a common political heritage almost entirely extraneous to the Alliance movement and the doctrines of greenbackism. The state level independent ticket lost narrowly, but in the three-party split the independents elected a majority to both houses of the state legislature. The fragility of the Nebraska reform movement quickly became apparent when the legislature convened in 1891. Despite their numerical majorities, the independents were able to advance only a truncated reform program centered around a weak railroad regulatory agency. The watered down legislation, known as the Newberry bill, finally passed, but it was vetoed by the Republican governor.

In the final analysis, the decade-long legacy of the Northwestern Alliance proved too much for the cooperative movement of the Alliance to overcome. The result in Nebraska was a fragile shadow movement unrelated to the doctrines of Populism that elsewhere had materialized within the cooperative movement. The agrarian cause in Nebraska had no institutional base, no collective identity, and no movement culture to counter the constant intimidation of the prevailing corporate culture. It possessed no mechanisms for self-education, no real lecturing system, no methods for developing individual self-respect among impoverished people. The farmer movement in Nebraska had no purpose. It only appeared to have one, because of its external resemblance to the real movement which did.

Whatever the Populist future was to hold, one thing was clear in 1890: the Nebraskans would not be on hand at the National Alliance convention in Florida with the organization credentials to assist their fellow Westerners from Kansas and the Dakotas in their self-assigned but politically decisive task of persuading the Southern Alliances to abandon the party of the Confederacy. In breaching the sectional barrier to achieve a multi-sectional third party, the Nebraskans were to prove of no value to the national reform cause.

6

None of which is to suggest that the high hopes of the spring had somehow been extinguished. On the contrary, the spectacular Kansas victories brought a surge of expectation to old Alliancemen from Dakota to Forida and to new Alliancemen from Colorado to Oregon. And, to many, another political event offered great promise—the sweeping Alliance conquest of the Democratic Party of the South.

To all appearances, the organized farmers of the Alliance achieved what appeared to be a "party revolution" across the South in 1890. In Tennessee the Alliance state president, John P. Buchanan, won the Democratic nomination and was elected governor. In Georgia the Democratic convention simply adopted the Alliance platform and nominated Allianceman William Northen for governor. The ensuing elections appeared decisive. Besides taking the Georgia governorship, Alliance-supported candidates won three-fourths of the seats in the state senate, four-fifths of the state house of representatives, and six of the state's ten congressional seats. "Being Democrats and in the majority," explained one Allianceman, "we took possession of the Democratic Party." Another exclaimed, "As in the day of Jackson, the people have come to power."

But there were some disturbing signs. In Alabama, where the Alliance claimed more than 75 of the 133 members of the state assembly, Democratic conservatives were able to prevail in the party's gubernatorial nominating convention, narrowly selecting the Bourbon leader, Thomas Jones, over the Alliance candidate, Reuben Kolb. In Tennessee the order's state president and new governor, in order to please what he regarded as his entire Alliance and non-Alliance constituency, decided it would be politically expedient to remain silent about most of the Alliance demands. And apparently while agricultural poverty might, under some conditions, lead to reform, it might also lead merely to personal political machines that traded on the language of reform. In South Carolina a ranting, one-eyed orator named Ben Tillman launched a campaign for power based on upland hostility to the tidewater gentry. He regarded the Alliance warily at first but eventually decided to join it, gain control, and use it for his own political ends. Though his hate-filled oratory terrified conservatives, the Tillman regime proved surprisingly amenable to the established order. Meanwhile in Georgia the victory of Governor Northen came to be seen more as a triumph for continuity than a breakthrough for the Alliance. Georgia newspapers regarded the new Governor as "progressive but safe."

Thus a close analysis of the 1890 elections might well have given the agrarian reformers reason to pause in their exultation. The party of the Confederacy had several well-fortified lines of defense. The party machinery remained in the hands of old-line regulars, and almost everywhere the controlling mechanisms of the parliamentary process—the chairmanships of powerful committees—as well as the leverage available through corporate lobbying influences, were retained by politicians oriented toward business rather than toward the Alliance. As of the end of 1890, however, one development seemed unarguable: by educating the farmers, the Alliance had brought the idea of reform into the dialogue of the Democratic Party of the South for the first time since the Civil War. Such, at any rate, was the way Alliance victories of 1890 appeared to the nation.

7

But from the viewpoint of Alliance radicals—and particularly from the viewpoint of the radical Alliance founders in Texas—

the cooperative movement had, ironically, accomplished precisely the opposite result. Far from being an energizing force, the cooperative crusade seemed to have sidetracked reform in Texas in 1888 and kept it there through the elections of 1890. As the old Cleburne radicals analyzed matters, the politics of 1890 in the South was a total sham, suggesting the mere appearance of reform without the slightest shred of its substance.

This interpretation, of course, differed markedly from that of other Southern Alliance leaders. Nevertheless, its origins demand attention, for its subsequent development charted the path that brought the Southern Alliances to the People's Party. The process generated something the American labor movement of the nineteenth century was never able to achieve: the politicization of masses of people into insurgent democratic politics.

The intellectual and tactical nucleus of Alliance radicalism consisted of six men who first came to prominence in the year of the Cleburne Demands in 1886 and who had subsequently acquired a degree of national prominence in the agrarian movement. Most visible, of course, was Evan Jones, the loquacious radical who successively served from 1884 to 1890 as chief spokesman for the movement in his home county, in his home state, and, in 1889, as president of the National Alliance. He had voted a third party ticket since 1884. A second insurgent tactician was H. S. P. Ashby, a member of the blue-ribbon panel in St. Louis that had brought forth the sub-treasury plan. Another prominent activist was J. M. Perdue, a greenback theoretician and principal author of the Cleburne Demands. The others were R. M. Humphrey, white founder of the Colored Farmers National Alliance; Harry Tracy, a sophisticated greenback advocate of large-scale cooperatives; and finally, of course, William Lamb.

In 1890 these spokesmen were a frustrated group. They were, perhaps, not as gloomy as Jerry Simpson of Kansas had been in 1889 after the Union Labor debacle, but they approached his despair and for the same reason: the failure of the third party effort both nationally and in their own state. And the height of the irony lay in the fact that the central cause of their difficulty was the cooperative movement itself.

The Texans had made a massive effort in 1888 to assist in the creation of a new national third party and to rally the entire Alliance to its support in Texas but their plans collided with the desperate struggle of the Texas Alliance to save its central state exchange. However necessary, Jones's decision to decline the third party gubernatorial nomination postponed for at least two years what he and his principal colleagues considered to be their central political mission—beginning the process of weaning rank-and-file Alliance farmers from their inherited loyalty to the "party of the fathers."

It became apparent in 1890 that the postponement might be for four years rather than two. As the emergence of Tillman in South Carolina and Northen in Georgia revealed, a new variety of Southern politician had materialized in response to the arrival of the Alliance in politics. In Texas the new aspirant's name was James Hogg, a 300-pound, railroad-baiting Democratic loyalist who asserted himself as a man of the people. Hogg had a colorful platform style that tended to obscure the lack of specifics in his reform program. While campaigning, he often took off his coat, threw his suspenders from his shoulders, letting them dangle about his knees, and drank "lik a horse" from a water pitcher. William Lamb, for one, was not impressed, particularly after reading the Democratic platform tailored by the Hoggites.

But in 1890 the old Cleburne radical had a weapon to employ that had been unavailable in 1888. Ironically, he could thank his rival, the "nonpartisan" Charles Macune, for the weapon was Macune's sub-treasury plan. The sweeping soft-money proposal was unlikely to please Democratic regulars, even when they displayed the trappings of reform in support of James Hogg. The Hoggites assisted in removing all doubt by resolving against the sub-treasury at the Democratic convention. But with no rival for the support of farmers, Hogg was safely "in" as the next governor of Texas. For radicals, the meaning of this event was as clear as it was ominous: the farmers of the Alliance may have become "conscious" of their economic relationship to bankers and credit merchants, but because of the continuing power of the received culture, they expressed their new-found sensibility by voting for old party candidates. They were, in short, not yet

"politically conscious." The agrarian movement in the South was
still unable to break the bonds of inherited political habits.

However, it was precisely at this point—August 1890—that
William Lamb set in motion the elaborate campaign of democratic education that was destined to transform the National
Farmers Alliance into the People's Party. First, he publicly
defined his opposition to the Southern Democracy. The Democratic platform had "sidetracked" the farmers, he said—particularly by the declaration against banks "without proposing a
substitute." Most outrageous of all, he added, was the pronouncement against the sub-treasury system, which was, of
course, the best available substitute. In the aftermath of the 1890
Democratic convention, Lamb readied his plan to drive this
home to the annual Alliance state meeting. The daylight between
the Alliance on the one hand and the Democratic Party on the
other had finally appeared in a way that farmers should be able
to see—for the sub-treasury was the one issue that addressed
the enduring realities in the lives of most Southern farmers, the
furnishing merchant and his crop lien. William Lamb proposed
to draw a radical distinction on the sub-treasury issue.

The arguments easily presented themselves. If the Alliance
was not willing to take a stand for the sub-treasury, then it
mocked its own claim that it represented the true interest of
farmers. Thirteen years of devoted experimentation with cooperative buying and selling had failed to alter the basic injustices
of the crop lien. Not only was this reality obvious to men like
Lamb and Macune, it was also apparent to everyone associated
with the cooperatives. The sub-treasury plan faced this harsh
fact squarely. Collectively, these arguments added up to a radical
ultimatum: if the farmers were unwilling to take on all comers
on the basis of the sub-treasury, they might as well abandon the
Alliance. "Cooperation" having failed to dislodge the furnishing
merchant, what other course was left? The Alliance had to stand
on the issue or concede its own irrelevance.

Four months before the Ocala convention of the National
Alliance Lamb pressed the choice on the Texas Alliance in the
order's state convention at Dallas. Lamb enjoined the Alliance
formally to endorse the issue that the Democratic Party had as

formally opposed. The pull of conflicting loyalties was evident, but awkward. Must one choose between the Democratic Party and the farmers? The supporting arguments for both sides— destined to be heard throughout the South in 1891–92—began to resound through Texas in 1890. Was not the Alliance strong enough to insist that the Democrats support the sub-treasury? The Democrats had already pronounced against it. Was not the Democratic Party the party of the people? If the party was not willing to try to cope with the furnishing merchant and the crop lien, it did not care about the people. What about the Democratic argument that the sub-treasury was class legislation, or unconstitutional, or both? Opposition to the sub-treasury was admission that the farmers could not be helped.

The arguments were stark and the choice painful. After a debate extending over two days, the roll call of counties was ordered. Sixteen counties abstained. Twenty-three voted "no." Seventy-five voted "yes." On a fundamental issue, the Texas Alliance had declared itself in opposition to the Texas Democratic Party.

But only on the leadership level. To the thousands of farmers in the suballiances, the relationship of candidates, political parties, and the sub-treasury was not nearly so clear as it had been to the Alliance leaders who debated Lamb's motion in August. The painful decision thrust on the Alliance leadership therefore had to be recapitulated at all levels of the order in a manner that would bring home to hundreds of thousands of farmers the choices that had to be made. To this course Lamb persuaded the agrarian leaders in Texas to dedicate themselves. Let the sub-treasury be explained by the lecturers. Let the farmers in the suballiances debate the plan. Let them see its implications in terms of their daily relationships with the furnishing merchant. Let the line be drawn.

Lamb mobilized the Alliance's internal organizational machinery to address an enormous lecturing task. The job was huge, for not only did the order's lecturers have to be increased in number, but they also had to be thoroughly briefed on the workings of the sub-treasury system and armed with intelligent defenses of its provisions. Lamb conceived of a plan to add an

additional layer of executive structure to the Alliance organization in order to meet the lecturing challenge. With the full
cooperation of Evan Jones, a multi-county lecturing school on
the sub-treasury was established in each congressional district.
County Alliance presidents, lecturers, and assistant lecturers
were convened and briefed, and speaking assignments were
integrated into a systematic plan to cover the hundreds of
suballiances in each district. In the fall and winter of 1890, one
congressional district after another organized its "lecturing
school." Lamb presided over the most elaborate one in November. By the end of the winter, the sub-treasury test that had
been put to the Alliance county leadership in August was being
recapitulated in hundreds of suballiances across the state. The
farmers were being asked to choose between the Farmers
Alliance and the Democratic Party. The Alliance had begun the
process of politicizing itself for a mass effort to redefine American
commerce and the American party system. The politics of
Populism had arrived in the rural districts of Texas.

The effort, however, was all uphill. While liberal journalists
in Washington might be writing optimistic interpretive stories
about "The Alliance Wedge in Congress," the Kansans knew
that wedge could easily be dislodged. The virulent attack on L.
L. Polk in Kansas had been an omen of what they might expect:
a party that seriously hoped to challenge the dominant Republicans in the North simply could not be tainted with the stain of
rebellion. Unfortunately for the activists, however, most Southern
Alliancemen seemed committed to "reform through the Democrats." The Southerners' position, moreover, was logically unassailable. They had created an "Alliance yardstick" composed
of the national demands formulated in the St. Louis Platform
plus a variety of local issues applicable in each state. They had
done so even before the Western Alliances had moved into
independent political action in the summer and fall of 1890.
Having announced their Alliance yardstick, the Southerners had
then measured candidates by it. So great was the Alliance sweep
that literally hundreds of legislators who had "measured up"
had been elected, and six Georgia Congressmen who had not
had been defeated. Indeed, the order promised to be in nu-

merical control of a number of "Alliance legislatures." To abandon the Alliance representatives even before their legislatures convened not only would make the order appear irresponsible before the world, it also would constitute an abandonment of the Alliance platform. How could the order go back to the voters as a third party with the same program without having given the men it had already elected a chance to act upon it? The question was unanswerable.

Still, when the National Alliance convened at Ocala, Florida, in December 1890, the Kansans pressed hard. Charles Macune provided a compromise. Though the new "Ocala Platform" was quite radical (formally adding the sub-treasury system to the list of "Alliance Demands"), Macune argued that a third party could service no one, neither the Westerners nor the Southerners, in 1891. "Reform through the Democrats"—of which there was no more fervent supporter than Macune—could be tested in the legislatures which would convene across the South in 1891. If that solution were found wanting, other alternatives, including a third party, could be considered early in 1892. Let the matter be postponed, said Macune. Agree at Ocala to reconvene on Washington's birthday in 1892. That date, only fourteen months away, would provide a much better view of the progress of the reform movement. As Macune put it, "If the people by delegates coming direct from them agree that a third party move is necessary, it need not be feared." The radicals accepted this solution—and then promptly went beyond it by calling a general reform convention at Cincinnati for 1891 in order to erect the preliminary organizational scaffolding of the new third party.

8

The decisive battleground, as Macune, Lamb, and the Western radicals all knew, was the South. It was also clear that if Southern farmers were to be persuaded to break with the party of the fathers, the Alliance founders would have to chart the route. Far more than the Kansans, the Texans possessed the essential sectional credentials to talk to other Southerners about abandoning the Democratic Party—credentials that were augmented by the fact the Texas lecturers had conducted the organizing sweep through the South in the first place. It seemed self-evident to the radicals in both regions that if the Southern Alliances were to join the third party, the Texas Alliance would have to lead the way.

William Lamb proposed to do just that by making the sub-treasury the fundamental test of Alliance loyalty. Under Lamb's formula, every farmer in the National Alliance was to be asked to stand up and be counted on the sub-treasury issue! The speed with which such a breathtaking assignment could be achieved depended on the strength of the Alliance organizational struc-ture in each Southern state as well as on the energy and political flexibility of its leadership. But first the South needed a model.

In the spring of 1891, radical Alliancemen in Texas expanded across the state the district lecturing schools on the sub-treasury pioneered by Lamb. Additionally, a legislative watchdog com-mittee was established in the state capital under the direction of Harry Tracy. Set up to render careful reports of the promised reform program of the incoming Hogg regime, the Tracy committee achieved surprising results in a matter of weeks. In a legislative atmosphere that they found to be crowded with railroad lobbyists, Tracy and the other Alliance committeemen soon lost their respect for "Hogg Democrats." The farmers got one of their state-level demands, a railroad regulatory commis-sion, but little else. After almost all of the anticipated reforms had been bungled, withdrawn, or postponed under relentless corporate lobbying pressure, Hogg's insistence that his legislative program was succeeding scarcely eased Harry Tracy's doubts about the governor's reforming zeal. Until that moment a loyal Democrat, Tracy broke with the party on the ground that Hogg's reform movement was devoid of substance. The chorus of anti-Hogg voices in the Alliance grew louder.

Governor Hogg, now deeply concerned, moved to discredit the Farmers Alliance. A "manifesto" attacking the Tracy legis-lative committee as well as the sub-treasury, signed by a half-dozen "Alliance legislators" allied with Hogg, received wide circulation in the state press. But Hogg had blundered: the attack came as the lecturing program on the sub-treasury was in full swing in suballiances across the state.

Amid mounting evidence that Lamb and the Alliance radicals were winning the Texas battlefield, Macune dropped matters in Washington and went home. He found himself more and more isolated in the Texas political environment. He had become a political rarity—a Democrat and a sub-treasury advocate. The contradictions implicit in his stance became more sharply clarified for him the longer he stayed in Texas. The crop-mortgaged farmers in the suballiances were discussing his plan with an intensity given no public question in the South since secession, but he could scarcely take comfort in the obvious meaning of such discussions. Everywhere he went in Alliance circles the sub-treasury had found increasing favor, but more often than not in the context of the need for a third party. After all, Jim Hogg was against the farmers on the sub-treasury!

9

Received cultures contain compelling political memories, however. Most Alliancemen in Texas had voted for "the party of the fathers" all their lives. Nevertheless, the farmers continued to make strides in their pursuit of an authentic democratic environment of their own; they no longer had to rely on vague reports in the business press to learn about their legislature or the other political institutions of the nation. In the increasingly democratic world they had created in their Alliance, they now had internal lecturing pipelines and their own watchdog committees. In this way, they were pursuing a new kind of political autonomy. In such a milieu—unknown to America since the Committees of Correspondence of the Revolution—self-described Democratic "reform" governors such as James Hogg and prestigious Republican Senators such as John Ingalls of Kansas were seen in a new and much altered perspective. This underlying development was the central one for the evolution of Populism: the people were making judgments derived from their own sources of information rather than accepting the interpretations of culturally sanctioned "leaders" as they had always previously done. In short, their new perceptions were in response to new democratic impulses rather than traditional hierarchical ones.

Yet it was not enough for "some" or "many" Alliancemen to attain this new plateau of individual self-respect. Given the hold the received hierarchical culture had on Americans outside the Alliance, the order needed to generate the new autonomy among all Alliancemen if democratic ways were to have a real prospect of prevailing in the larger society. On this fundamental level of self-determination, all Alliancemen had to see the need to rely on themselves individually, and on each other, rather than passively acting out received habits of political conduct. Now, more even than on "the day to save the exchange" in 1888, the farmers needed to talk to each other and chart a democratic course for themselves and for the nation. Clearly, in the spring of 1891 a crucial test of the democratic aspirations of the Farmers Alliance was at hand.

The agrarian leadership moved decisively. In order to expand the already enlarged district lecturing system on the sub-treasury, Evan Jones issued a proclamation calling the "first annual Alliance Conference"—a special four-day statewide "educational" meeting in Waco on April 21, 1891—for the express purpose of "perfecting the lecturing system." Macune could not oppose such a program, but he had read enough signs to take precautions. He arranged to have well-known national Alliance Democrats invited to the conference. Yet the spread of third party sentiment was evidenced by other guest speakers invited to Waco, all recruited from the membership rolls of the National Reform Press Association: Henry Vincent and M. L. Wilkins of the *American Nonconformist* and Ralph Beaumont of the *National Citizens Alliance* of Washington. The tactical importance of the Texas Alliance was now well understood by both factions of reformers throughout the nation. As for the sub-treasury, both were, of course, fully committed: on the eve of the Waco meeting, that was the one certainty accruing from years of Alliance experience with the oppressive features of the American monetary system.

This latter circumstance was lost on the metropolitan press and on Hogg himself. Both were far removed from the economic hardship that undergirded the agrarian revolt. From the comfortable perspective of Democratic politicians and the major

Texas dailies, a battle was expected at the Alliance meeting—but it was presumed to concern Alliancemen dedicated to Hogg and Alliancemen loyal to Macune and the sub-treasury plan. In hierarchical societies, genuine democratic politics, when it appears, is hard to understand.

It was with some dismay, then, that reporters discovered during the first two days of the Waco meeting that the Texas Alliance was "united" behind its "official family." To the eyes of the press, this included Macune, Evan Jones, Lamb, Perdue, Tracy, and Ashby. In general, the Waco meeting seemed to border on a mass celebration. "Stump" Ashby was "the favorite," the "famous agitator and humorist," while Evan Jones, "sadly needing a patch on his pants," was "by odds the most popular leader." Taken together, the Alliance "official family" was "now in the heyday of power and popularity," while the half-dozen or so Hoggites among the 400 Alliance county leaders were "discouraged."

But this solidarity existed only when the Alliance leadership was described in terms of personalities rather than policies. The truth was that esteem and policy-making had, in the case of Charles Macune, ceased to be mutually supporting elements. Though the "nonpartisan" *Economist* editor was the best known Allianceman in America, he publicly revealed his inability to enunciate a clear policy for the order. In the name of the sub-treasury, he stood for legislative political action, though not for the only method through which it could be obtained—independent political action. Because both major parties had rejected the sub-treasury, the politics of the issue left Alliance "nonpartisans" such as Macune defenseless against the simple political logic of the order's radicals. This situation became clear in Texas in April 1891, as it was to become clear in Georgia in the spring of 1892 and in other parts of the South later in the same year. Macune's words at Waco revealed both the extreme delicacy of his position and his uncertainty about his own future course.

> I remember that five years ago, the Alliance was afraid of politics, but the order has got bravely over that . . . and if it is necessary as a method of accomplishing their aims to enter

the ranks of politics and dislodge some of the leaders herea-
bouts, they are equal to the emergency. . . . I have learned a
great deal since joining the order and expect to learn more.
I felt when I first joined I could give a better description of
its objects than I can today. . . . I am not afraid of politics.
. . . But . . . let us use it as a method, never as an object.

Those were no longer the words of a man who was leading.
In sudden increments of discovery, the Waco meeting materialized before observing reporters as a series of shocks, each
more startling than the last.

The pace of events was rapid. On the first day Lamb attempted
to engender leadership agreement to dispatch a delegation to
Cincinnati; it ran afoul of Macune's contention that the Alliance,
as an organization, could not commit itself officially to a specific
party. On the second day a compromise was reached to permit
the decision to be made by individual caucuses in the congressional districts. The shifting center of gravity became still clearer
on the third day of the conference, when the Alliance radicals
narrowly lost a series of indirect test votes on the third party
issue by margins of 85 to 83 and 82 to 81. Debate was not heated,
as the delegates were intent on appraising the political pulse of
the order. Another compromise was quietly achieved. The
radicals agreed to confine further third party exhortation to
separate meetings already scheduled in Waco; in exchange, the
Texas Alliance would send delegates to Cincinnati. The same
day, before "large and enthusiastic" crowds of third-party advocates in the Waco courthouse, two new institutions were
created—the Texas Citizens Alliance and the Texas Reform
Press Association. The press association's new president was
William Lamb and its vice president was to be the editor of the
Southern Mercury, the official journal of the State Alliance. The
Texas Citizens Alliance was organized as an affiliate of the
overtly third party National Citizens Alliance, and *its* secretary-
treasurer was William Lamb.

Taken together, these events produced a new view of the
meaning of the "first annual Farmers Alliance conference." The
Waco *News* confessed it had found the conference truly "educational" and, in reporting on the county presidents and lecturers, said that "one is forced to suspect that they have been to

school before." "Enough has been learned," the paper added, to conclude that the meeting "has a political significance of no mean importance" and that "the third party is in fact, a probability."

Lamb had won the tactical war with Macune. In conceiving the political tactics of the sub-treasury, Lamb had integrated the lecturing system and the reform press into an effective instrument of political democracy. As events were soon to show, the "politics of the sub-treasury" was the critical tactical instrument needed in the South to "draw the line" between the farmers' loyalty to their Alliance and their inherited cultural loyalty to "the party of the fathers." It made them confront the reality of the credit system in ways that had tangible political meaning. In this manner, Lamb provided the South with a model through which the Alliance could be brought to Populism.

Yet there was a much deeper meaning to the events at Waco. The "first annual Alliance Conference" revealed how William Lamb, Charles Macune, and Alliancemen generally had been lifted by their collective cooperative efforts to a new plateau of democratic possibility. All of the collective acts of the Alliance— the wagon trains, the encampments, the many modes of cooperative striving—were practical demonstrations of group self-confidence. But there was something else: cloth flags flying from wagons heading home after the successful bulking of one's crop, a private celebration. And at the collective celebrations, at the twilight meals for thousands, if, as inevitably happened, there came speeches about the "new day for the industrial millions," followed by "cheers for the Speaker and for the Alliance," these were also cheers for oneself, for one's new vision of hope that had come to life in the democratic environment of the Alliance.

The link that connected the people of the Alliance, that carried the hopes of the many forward to their elected leaders and carried the response of the same leaders back to the brotherhood, was the Alliance lecturing system. Here lay the essential democratic communications network within the movement and it was this ingredient that Lamb mobilized in the spring of 1891 for yet one more post-Civil War attempt to bring a democratic "new day" to America. Through the politics of the sub-treasury, with its vividly direct connection to the crop lien system of the South, Lamb put the movement culture that had grown up out of the cooperative crusade to its ultimate test against the received culture of inherited sectional loyalty to "the party of the fathers."

Irony suffused the struggle between Lamb and Macune. On the one hand, their battle over Alliance policy marked the intersection of their influence in the agrarian crusade. For the time thereafter available to both, Lamb's impact on policy increased while Macune's declined, as the radicalism Lamb had engendered at Cleburne in 1886 continued to transform the national organization Macune had envisioned at Waco in 1887. But though their long ideological conflict was not finally to culminate for another nineteen months, their debts to each other were already plain in April 1891 and could be summarized in three facts: Lamb's radicalism, splitting the Alliance founders in 1886, brought Macune to power; Macune's organizational creativity constructed a national constituency of farmers that made possible Lamb's dream of a third party of the laboring classes; and their shared objective, economic parity for farmers in an industrial society, was symbolized in the sub-treasury plan, which Macune conceived as an instrument of economic reform and Lamb converted into one of political revolt. In the process, the stakes were substantially raised. The struggle to free the Southern farmer from the furnishing merchant was subsumed in the struggle to bring economic redress to "the industrial millions." It was a contest Macune felt could not be won, and one Lamb felt had to be waged. The respective successes of Lamb and Macune and the partisan attacks those successes inspired against each, plus their own actions, eventually used up the political credit of both. A third casualty was the sub-treasury plan. The achievement was the creation of a multi-sectional institution of reform: the People's Party. Macune's sub-treasury, in Lamb's hands, defeated Macune and created Southern Populism.

10

In the state conventions of the Alliance across the South in the summer and fall of 1891, Alliance leaders were asked to stand up and be counted on the sub-treasury. The political cost to the Alliance was sometimes substantial, but at whatever cost, the line was drawn. As Lamb had demonstrated in Texas, it marked out the starting point for the People's Party in the South. However painfully—and it soon became quite painful in some states—the politics of the sub-treasury, fashioned some twenty-six years after Appomattox, became the sword that cut the ancestral bonds to the party of the fathers.

Third party partisans got some help from the business interests of the South. Throughout 1891, as Southern state legislatures were put to the test on the "Alliance yardstick" of legislative reform, the business orientation of the party of the fathers revealed itself. One by one, the legislatures of the Southern states did not produce "reform through the Democrats." Whether Democratic officeholders styled themselves as traditional "states' rights" conservatives or as "Alliance Democrats," the intellectual and political distance between themselves and the farmers became all too evident.

What happened to the Hogg "reform program" in Texas happened everywhere. The American political system was not seen to be democratic, but hierarchical; business lobbies governed the legislative process on the vital issues. In 1891, as it had many times before, the Democratic "party of the people" revealed itself as a business party.

In general, the Southern situation provided promising soil for the sub-treasury. Sometimes orchestrated through specially constituted lecturing systems organized by congressional districts, and sometimes not, the politics of the sub-treasury began to penetrate downward to the thousands of suballiances across the South in 1891. Beginning in the summer and extending into the following spring, state, county, and local alliances formally declared for the sub-treasury as the essential element of the "full Ocala Platform." The battle for the political allegiance of the Southern yeomanry was on.

Nowhere did the radical imperative to break the loyalty of Southern farmers to "the party of the fathers produce a more tense and vivid political struggle than in Georgia. The sprawling Georgia Alliance became a dramatic and bitter battleground as the state's press and politicians watched anxiously and, wherever possible, added their influence on the side of orthodoxy. To many observers, the struggle for the soul of the Georgia Alliance appeared to be a war between two rival captains—Lon Livingston, Alliance state president, and Tom Watson, radical agitator. Though this assessment has a measure of truth, Livingston and Watson were symbols of a decision-making process that ultimately had to find its resolution through the individual choices of thousands of Georgia farmers.

Near the end of his life Watson fondly recalled the "radiant visions" of the years of struggle in the 1890's: "I did not lead the Alliance; I followed the Alliance, and I am proud that I did follow it." Yet the relationship of the movement to the movement's spokesman was necessarily mutually supportive. For his part, Watson knew that the crop lien, and the systems of politics and justice that supported it, was the source both of the despair of the Georgia countryside and of the hope represented by the Alliance. "Here is a tenant—I do not know, or care, whether he is white or black, I know his story. . . . He knows what an order to the store means. He knows perfectly well that he cannot get goods as cheap as for cash." The system "tears a tenant from his family and puts him in chains and stripes because he sells cotton for something to eat and leaves his rent unpaid." When such a system could not punish its "railroad kings," it was "weak unto rottenness."

Yet beyond the language of reform, Watson early understood that reform movements require tactics and strategy. Even as Livingston signaled clearly his intention of staying with "the party of the fathers," Watson fomented properly radical preconditions for the decisive choices yet to be made in 2000 Georgia suballiances. It took a while for such a process to work. Livingston's position within the Alliance appeared invulnerable. He knew the rhetoric of revolt and could hurl thunderbolts at railroad barons with every bit of the fervor of a Tillman in South Carolina or a Hogg in Texas. He could avoid the taint of conservative party leaders by questioning their actions, as he did Governor Northen's, meanwhile following a conservative strategy of Democratic Party loyalty. The defenders of the received political culture had many weapons.

It was from this position of relative isolation that Tom Watson renewed his struggle. He began to put renewed emphasis on the sub-treasury plan and got himself a newspaper to carry the message. Watson's journal first appeared on October 1, 1891, as *The People's Party Paper* even as the politics of the sub-treasury became the center of attention in the Georgia suballiances.

Like *The People's Party Paper* in Georgia, L. L. Polk's *Progressive Farmer* in North Carolina focused increasingly on the sub-treasury as "the one real living issue," while Alabama reform editors, supported by the state Alliance president, also flocked to the sub-treasury standard. "The people are learning that the sub-treasury means final freedom from serfdom," one editor put it. A farmer punctuated the judgment: "Hurrah for Ocala, first, last, and forever! Amen." In South Carolina, too, embattled Alliance radicals seized upon the sub-treasury in a desperate effort to "draw the line" between the South Carolina Alliance and Ben Tillman.

To outsiders the full implications of these furious controversies were not always much better understood than when the Texas Alliance took the same step in the summer of 1890. But the same internal dynamics were at work. To all who looked, the politics of the sub-treasury measured the strength, reform instinct, and internal cohesion of the Alliance movement in each of the Southern states. The votes varied from near unanimity among the county leaders of the Florida and Arkansas Alliances to an ominous postponement in Tennessee, where the issue came at an awkward moment in the political life of John Buchanan, the Alliance state president. The difficulty in Tennessee centered on the fact that Democrat Buchanan also occupied the gubernatorial chair, a circumstance that made him pause before launching a tactical campaign to draw a line between the Alliance on the one hand and the Democratic Party on the other. After much behind-the-scenes maneuvering, the Tennessee Alliance trimmed its sails on the Ocala Platform and postponed any and all tests of sentiment until 1892. The decision effectively destroyed the third party movement in the state.

The central role of the sub-treasury debate as the capstone of political education within the movement culture of the Alliance—

and the equally central need for farmers to have enough time for this process of democratic self-education to take place—was made dramatically clear in Mississippi. There, the politics of the sub-treasury erupted amid the 1891 state elections. Alliance leader Frank Burkitt was forced to throw the sub-treasury into battle against the state's planter Democrats even as a district lecturing system was mobilized for internal "Alliance education" on the same issue. The pace of events forced the Mississippi Alliance to outline the intricacies of the sub-treasury to Mississippi farmers as a part of the campaign itself. Though stump oratory was hardly the best medium for dissecting an issue as unusual and controversial as the sub-treasury, "immense crowds" took part in the canvass. The farmers of Mississippi, wracked by racial phobias but wracked also by a generation of the crop lien, were plainly in a volatile state.

The press of Mississippi became stridently engaged against the sub-treasury. Burkitt, the state lecturer of the Mississippi Alliance, took to the campaign trail and showed the order's county lecturers how to put the politics of the sub-treasury to work in the reform cause. In bone-poor Mississippi, Burkitt clearly scored telling points. By the early fall observers privately agreed that the Mississippi electorate, centered in the huge (60,000-member) state Alliance, could not be counted in anyone's camp. A new kind of Southern politics had come to Mississippi.

In October an older politics reasserted itself. Night riders descended on Frank Burkitt's *Choctaw County Messenger*. They set the building afire and destroyed the printing press. The courthouse at Pontotoc was broken into and all the voter registration books were "stolen or concealed or probably burnt." The mood of the campaign altered. The canvass became "one of the meanest ever conducted in the state," and the audiences became quieter. Mississippians were thinking things over, but since they no longer talked as freely one could not be certain of the situation. On election day, the regular Mississippi Democracy won a clear-cut victory.

Party regulars, aware that the returns were misleading, privately conceded that the election had been a near thing. The old order had repelled the first attack, but the margin of victory—and perhaps even more—had been a product of the final stages of the campaign. Aside from a ramshackle political apparatus that involved state and county officeholders, the old party's hard-core support had come from the furnishing merchants, many of whom had become large landowners over the years. The great mass of Mississippi people remained poor farmers, a fact the Democracy belatedly focused upon. The party's public appearance, so long neglected, definitely required face-lifting for any future engagements with its impoverished electorate—perhaps something along the lines Hogg had followed in Texas, or something resembling the Tillman regime in South Carolina. Such were the post-election imperatives facing Mississippi's Democratic chieftains. From the perspective of Alliance leaders, the postmortem conclusion was much simpler: it had all happened too quickly. They had not had enough time.

In contrast, in Georgia, where matters proceeded at a more orderly pace, four of every five suballiances were endorsing the sub-treasury. Still, matters remained uncertain everywhere. Abandon the party of the fathers? The party of the Confederacy? The party of the white man? Across the South, independent political action represented an agonizing cultural decision. Alliancemen, made economically conscious by the lessons learned in their cooperative crusade, were being asked to graduate to a new plateau of political consciousness through the continent-wide debate on the sub-treasury system. Slowly, then with accelerating speed, the conversions began through every echelon of the order. Indeed, the most notable recruit was the Alliance national president.

11

L. L. Polk's pace toward the third party, as indicated by his 1890 performance in Kansas, had quickened almost from the moment of his election as president. Though his decision to refrain from a public endorsement at the time of the Cincinnati meeting in May 1891 had not been well received by Alliance radicals, he privately conveyed a message to his son-in-law: "Let 'em rage, I will come in on the home stretch." By the time of the annual year-end national Alliance convention, held in Indianapolis in 1891, the North Carolinian had found his "home stretch." He advised the farmer leaders "to be deceived no longer" by "arrogant party dictation" and provided a demonstration of how a united third party might parry the blow of sectional agitators in both parties.

> To the charge that our organization is dominated by Southern influence, we have only to call the roll of this body to find that of the thirty-four states comprising it, twenty-three of them are denominated Northern states. . . . Not the war of twenty-five years ago . . . but the gigantic struggle of today between the classes and the masses . . . is the supreme incentive and object of this great political revolution.

Polk's address, which drew "thunderous ovations" from Alliance delegates gathered from across the nation, symbolized much of what the agrarian revolt had been since the days of the Cleburne Demands in 1886—Southern in origin, national in purpose, radical in ideology. As one newspaper reporter put it after Polk finished, the election of anyone else as Alliance president for the coming year of struggle "would have been regarded as a blow to the People's Party." Populist representatives, led by H. E. Taubeneck, the provisional national chairman named at Cincinnati in May, were "happy as clams." So was Tom Watson. *The People's Party Paper* declared that "Georgia is ready for a third party and will sweep the state with the movement." In this way, the Alliance in 1891–92, through its self-generated movement culture, politicized itself for democratic acts.*

---

* The venerable Marxist term, "class consciousness," is simply too grand an abstraction to serve coherently as a precise description of the complex process through which insurgent democratic movements (1) form, (2) recruit a mass base, (3) achieve a heretofore culturally unsanctioned level of economic analysis, and (4) find a way to express this new economic understanding through autonomous political acts. As the industrial tumult of the 1880's and 1890's indicated, American workers were increasingly "class conscious" in the traditional economic meaning of the phrase. But, as evidenced by continuing working class support of major party candidates, they found no autonomous means to express this consciousness in practical political ways.

In terms of the agrarian revolt in America, the four stages cited above were sequentially attained not merely through "class consciousness," but as a result of the development of (1) individual self-respect, beginning with Daws, Lamb,

12

All across the Northwest in 1890–91, however, the Northwestern Alliance, having never developed a movement culture, slowly disappeared from the agrarian revolt.† In Minnesota, the sheer momentum of the Loucks-Wardall-Donnelly cooperative campaign had seemed to carry the order to new heights of membership, but the departure of Loucks back to South Dakota and of Wardall to a new role in Macune's Washington office removed from the Minnesota leadership the only people who understood the organizational principles of the cooperative movement. Donnelly continued to believe that "the masses" could be organized by stump speeches and policy formulated by parliamentary dexterity in the nation's legislatures. Through simple neglect and organizational innocence, the fragile framework of a mass movement set loosely in place in Minnesota gradually, almost gracefully, fell apart in 1891. Donnelly continued to see politics from the top down as his role in platform-writing for the national People's Party brough him a modicum of national attention. He remained optimistic, unaware that he led a movement in his home state that possessed a steadily decreasing number of followers. "The sky is luminous with promise," he said in 1891. The inevitable electoral shocks of November 1892 in Minnesota were to leave him baffled and embittered.

Elsewhere in the territory of the Northwestern group, the Iowa Alliance, under the conservative direction of loyal Republican state presidents, never developed an internal program of

Vincent, Loucks, Watson, *et al.;* (2) collective self-confidence initially attained through the successful bulking of cotton, the war against the jute trust, etc.; (3) the economic lessons learned as a result of banker and merchant opposition to the cooperative movement as a whole; and (4) the creation of a mass-based third party through the politics of the sub-treasury as disseminated by the Alliance's internal lecturing system.

Though the precise pattern will naturally vary from country to country and movement to movement, collectively these sequential stages that produce mass democratic insurgency yield a mode of conduct antithetical to the social, economic, and political values of the received hierarchical culture. For the agrarian revolt in America, I have called this sequential development "the movement culture." It can correctly go by another name: Populism.

† What follows here, and on pp. 174–83, is essentially a brief sketch of the process by which an incipiently insurgent movement fails (1) to recruit a mass base, or (2) to politicize its members, or (3) both. See also pp. 139–44.

interest to the state's farmers. "The Farmers Alliance is in the worst shape imaginable," wrote one member just before he decamped to the National Alliance. In the same months of 1891, Henry Loucks also became disillusioned: "I regret the Alliance in Iowa is so backward in taking hold of our economic measures," he wrote. L. L. Polk, equally dismayed, ordered an organizing assault by the National Alliance on the Iowa territory of the Northwestern group. But simply too little foundation existed upon which to build a statewide cooperative movement. Though the farmers of Iowa suffered from the same financial and marketing practices that plagued farmers elsewhere, the doctrines of cooperative-greenbackism remained alien to them. Politically, therefore, they did not know what to do about their plight.

The months of 1891 saw the slow decline of Jay Burrows's political status in Nebraska and a corresponding rise in the influence of John Powers who endeavored to work congenially with incoming third party men. After wholesale membership losses in the Alliance in 1891, Powers had finally grasped the relationship of the cooperative movement to organizational continuity. He moved to put some flesh on the ramshackle Nebraska lecturing system, noting rather sadly at the state convention that "there seems to be a disposition in some of our subordinate Alliances to turn their meetings into mere literary entertainments or debates." The National Alliance made one final effort to activate the Nebraskans in January 1892, this time successfully. In another sharp departure from preceding years, Powers also stressed the sub-treasury plan in his presidential address, and the Nebraskans dutifully tracked the platform of the National Alliance. But coming as late as they did, these moves had only marginal impact on the disorganized structure both of the Alliance and the third party in Nebraska.* They had even less effect on the organizational residue still clinging to life in other states of the Northwest. The "national" meeting of the Northwestern Alliance, held in Chicago late in January 1892,

---

* In contrast, the Alliances of Kansas and the Dakotas engendered elaborate lecturing campaigns on the sub-treasury. Goodwyn, *Democratic Promise*, pp. 649–52, fns. 34 and 39.

bared the terminal illness of the order. Though Ignatius Donnelly and a number of Nebraska delegates arrived to argue in behalf of independent political action, the Northwesterners flatly refused to participate in the 1892 Populist national conventions and formally dissociated themselves from any connection with the third party movement. The Northwestern order, its members sharing a penchant for meeting in congenial rural settings and little else, numbered fewer than 25,000 in its entire jurisdiction. After the Chicago meeting, the Northwestern Farmers Alliance was almost literally never heard from again.

13

By the time of the long-awaited general conference of reformers, due to convene in St. Louis in February 1892 under the Macune formula at Ocala fourteen months earlier, the politics of the sub-treasury had stripped the plan's original proponent of all persuasive arguments against the third party. At St. Louis Macune maintained a low public silhouette, while L. L. Polk, having entered the "home stretch" at Indianapolis two months earlier, drove toward the finish. In an address of welcome at St. Louis that drew rising ovations, Polk said: "The time has arrived for the great West, the great South, and the great Northwest, to link their hands and hearts together and march to the ballot box and take possession of the government, restore it to the principles of our fathers, and run it in the interest of the people." Polk's election as the convention's chairman ensured the immediate formation of the People's Party. But the emotional peak at St. Louis was provided by Ignatius Donnelly of Minnesota. Donnelly's famous preamble was an expression of the deepest drives of the agrarian radicals who filled the hall and who had worked so many years to gain the allies who joined them there. If Donnelly's words seemed harsh and excessive to the comfortable, the delegates at St. Louis felt he described the American reality:

> We meet in the midst of a nation brought to the verge of moral, political and material ruin. Corruption dominates the ballot box, the legislatures, the Congress, and touches even

the ermine of the bench. The people are demoralized. . . .
The newspapers are subsidized or muzzled; public opinion
silenced; business prostrate, our homes covered with mortgages, labor impoverished, and the land concentrating in the
hands of capitalists. The urban workmen are denied the right
of organization for self-protection; imported pauperized labor
beats down their wages; a hireling standing army, unrecognized
by our laws, is established to shoot them down, and they are
rapidly disintegrating to European conditions. The fruits of
the toil of millions are boldly stolen to build up colossal
fortunes, unprecedented in the history of the world, while
their possessors despise the republic and endanger liberty.
. . . We charge that the controlling influences dominating the
old political parties have allowed the existing dreadful conditions to develop without serious effort to restrain or prevent
them. They have agreed together to ignore in the coming
campaign evey issue but one. They propose to drown the cries
of a plundered people with the uproar of a sham battle over
the tariff, so that corporations, national banks, rings, trusts,
"watered stocks," the demonetization of silver, and the oppression of usurers, may all be lost sight of. . . .

Men and women surged forward to surround Donnelly on
the platform and grasp his hand. Waves of enthusiasm greeted
the presentation of the Populist platform itself, which, being
shorter than Donnelly's preamble, was quickly read. A date for
the party's first presidential nominating convention was set: July
4, 1892.

As the convention raced toward adjournment, a defeated but
thoughtful Charles W. Macune silently decided to follow his
National Alliance into the third party. He was not one to remain
a follower for long. As the gavel signaling adjournment fell,
Macune leaped to the stage and shouted to the delegates to halt
in place and return to their seats. Startled, they obeyed. Carefully,
Macune explained that the Ocala convention had established
machinery to plan the meeting they had just attended, in
conjunction with others of like mind who also desired a grand
amalgamation of the laboring classes. Now that the convention
was ending, new machinery had to be established so that orderly
planning might ensue. With the delegates responding attentively

to his admonitions from the speaker's podium, Macune had become once again—for a moment, at least—the presiding captain of the reform movement. As the officials of the People's Party listened with what can be surmised were mixed emotions at best—the provisional national committee included, among others, William Lamb—Macune suggested that the convention reassemble immediately as a committee of the whole to establish its third party administrative apparatus. The delegates agreed and one of the fifteen named to this committee, in the nick of time, was Macune. His ambivalence on the third party issue forfeited any possible claims to popular leadership of the movement. The new party's presidential nomination would go to L. L. Polk, not to Macune. Nevertheless Macune was still editor of the movement's national journal, and his considerable organizing talents and diplomatic skills might enable him to become a successful and, perhaps, honored party chairman. But the convention's committee of fifteen soon found itself absorbed by the national committee and no one in either group rose to suggest that Macune, rather than the newly appointed Taubeneck, should head the crusade into its political phase. So much work loomed for everyone, in any case, that questions of leadership awaited future consideration. These ad hoc arrangements were to have large consequences for the future of Populism, for Taubeneck was to become a central character in the "free silver" controversy in 1895–96.

The chief anxiety among third-party strategists in the four months between the two inaugural Populist conventions concerned the progress of the campaign by Southern radicals to wean the Southern Alliances from the party of the fathers. Given the fate of all third party efforts since the Civil War, the concern about the South outweighed all other considerations.

As the Southern Alliances continued to employ their internal lecturing systems on the sub-treasury after St. Louis, L. L. Polk called a South-wide conference for Birmingham in May. But his purposes were not understood in the West, and Stephen McLallin, editor of the official Alliance newspaper in Kansas, wrote Polk a pleading letter that starkly revealed the continuing impact of sectional politics on radical hopes.

> The call for meeting at Birmingham . . . is causing considerable
> anxiety here. Your signature to the call alone allays suspicion
> with regard to the purpose of the meeting. . . . You know our
> people are extremely nervous at this time, and while they are
> beginning to believe that the south is with them, they are
> trembling lest something may happen to disappoint them. I
> cannot tell you how anxious they are or how they are hoping
> almost against hope that this momentous year will show them
> "the way out." You know these things better than I can tell
> them to you. I give you the fact that this call is causing much
> discussing and anxiety here and you will exercise your own
> judgment as to what it is best to say, if anything, in regard to
> it.

The Kansans had done all they could to help. Whenever they
could spare the time from their own political movement in
Kansas, Henry and Cuthbert Vincent, Jerry Simpson, Mary
Lease, William Peffer, and others toured the South to provide
visible evidence that non-Southerners had made the break with
political tradition, too. Indeed, the willingness of the Kansans
to expend such energies provided a measure both of their
concern and their need.

Polk's concern matched those of his Western allies. In Birmingham the Alliance president engaged in the delicate business
of holding together a "nonpartisan" order as it moved into
independent political action. The two-day conference of thirty-
seven top Alliance officials from eleven Southern states endured
intermittent tension, but ended on a relatively amicable note
addressed specifically to "the brotherhood in the North and
great Northwest." The Southerners announced their intention
to work "in unison," to "stand by them in all laudable efforts to
redeem this country from the clutches of organized capital,"
and, finally, to "stand with them at the ballot box for the
enforcement of our demands." Though the Alliance statement
was signed by all thirty-seven participants, the delicacy of the
situation was underscored by an additional fact not included in
the press announcement: despite Polk's importunings, the Southern Alliance leaders declined, by a vote of 21 to 16, publicly to
express an outright endorsement of the People's Party. Ten-

nessee and Florida, both with prominent "Alliance Democrats" high in their counsels, were balking and Ben Tillman was working zealously to clamp a lid on insurgency in South Carolina. Even in North Carolina, young Marion Butler, who had replaced Polk as state leader when the latter had assumed the order's national presidency, had fashioned an elaborate scheme that involved cooperation with North Carolina Democrats in state races while charting a third party course in national politics. Clearly, the radical agrarian leadership in the South had much work still to perform. The cause of independent political action in the states of the Old Confederacy still rested on shaky grounds. Polk's energies, skill, and prestige obviously confronted an ultimate test.

Meanwhile, throughout the nation, radicals held their precinct, county, and state conventions to select the delegates who would go to Omaha to convene the new People's Party of the United States. The selection procedures worked unevenly, needless to say, but by one method or another, through democratic processes or through self-selection, some 1400 certified Populists appeared in Omaha on July 4, 1892, to nominate the national standard-bearer of the "new party of the industrial millions."

14

L. L. Polk was not among them, however. Three weeks before the convention, on June 11 in Washington, the fifty-five-year-old Alliance chieftain died suddenly, after a brief illness. His loss altered the thrust of Populism—how much will never be known—but enough, certainly, to camouflage for many subsequent students the movement's strength across the South. At least one Boston observer of the national political scene, H. H. Boyce of the liberal journal *Arena,* had told Polk just before his death that he was "the one man in the country who can break the Solid South." The judgment was well-considered. Through his personal standing and inherent appeal in the South, Polk's candidacy might have done much to make up for the extremely late start the third party attained in 1892 through most of the Old Confederacy.

Beyond this, it is difficult to imagine a third party leader more uniquely armed to counter the politics of sectionalism than Polk, the old Southern Unionist from North Carolina. His performance, from New York to California, on his exhaustive speaking tours of 1890–91 forcefully demonstrated that his ability to transcend the sectional barrier was as highly developed as that of any other politician in the nation. His untimely death was a heavy and unexpected blow to Populism in the South and, conceivably, elsewhere.

The new party was fated to encounter sectional prejudice whomever it nominated. With Polk heading the ticket, the West would have had to bear the first shock of sectionalism as Republicans pressed "bloody shirt" attacks against the third party's "ex-Rebel" leader. As it was, the nomination at Omaha of James Baird Weaver elevated a former Union general from Iowa to the head of the new party's ticket. This decision ensured that the new leaders in the South would be the ones to receive the first wave of sectional attacks. The Weaver ticket was balanced with Virginia's James G. Field, giving the new party both an ex-Unionist and an ex-Confederate at the head of its ticket. The party they sought to lead had a platform that addressed the "living issues," a preamble that excited its true believers, and a structure of state leadership at the top, but the most direct initial appeal the new party could make to masses of people was in the form of a memory—the emotional and recent memory of the Farmers Alliance.

The cooperative crusade had brought hope to millions, a victory over the jute trust, and a provocative political upset in Kansas. Had it done more? What was the true significance of the long trains of Alliance wagons in Kansas, the huge encampments in Texas, the growing adoration that farmers, their wives, and their children bestowed on a fiery young Congressman named Tom Watson? What did it all mean? In Omaha 4000 delegates and spectators had erupted in sustained cheering, clapping, yelling and crying after hearing, one by one, the political planks of the Alliance "demands" they would come to revere as "the Omaha Platform." To one observer who saw in it the frightening specter of socialism, the sounds of the convention "rose like a tornado" and "raged without cessation for thirty-four minutes" while "men embraced and kissed their neighbors, locked arms, marched back and forth, and leaped about tables and chairs." Was the People's Party a hope or a threat? To *The New York Times* the sub-treasury plan appeared as "one of the wildest and most fantastic projects ever seriously proposed by sober man," yet alongside the rolling thunder of Ignatius Donnelly's preamble Macune's formula seemed as thoughtful and attainable as its author had always insisted it was. Could such seeming variety have materialized solely from the Alliance experience, so recent and yet so impassioned? What was shadow, and what was substance?

The men and women of the Omaha convention were asking their new party to overcome deeply ingrained sectional, religious, and racial loyalties in the name of their vision of reform. Conceivably, the biggest obstacle facing the People's Party might be the culture of America itself. But such a thought would not have traveled far among the ranks of the Populist faithful: the one current that merged the Alliance with the People's Party, and Northerners with Southerners, was their shared faith: though the democratic heritage was imperiled by the demands of the industrial age, the people were not yet helpless victims. They could, in Polk's words, "link their hands and hearts together and march to the ballot box and take possession of the government, restore it to the principles of our fathers, and run it in the interest of the people."

Across the Western granary and the Southern cotton belt, the People's Party would test the depth of the democratic culture that was the new party's inheritance from the Alliance cooperative crusade. Earnest radicals fervently prayed that the huge Alliance constituencies had learned their greenback lessons well and that the knowledge would hold them steadily on an independent political course in the face of all the cultural counterattacks certain to appear. The nation, in any case, would soon know, for the "coming great contest" had, indeed, come.

6

Reform Politicians,
Reform Editors,
and Plain People
The Language of American Populism

"Welcome honorable allies and go forth to victory."

In the absence of such highly charged stimuli as a revolution or a civil war—events that create massive political constituencies overnight—new institutions such as the People's Party necessarily face the prospect of building out of the material at hand. The National Farmers Alliance was to convey masses of rural people to the new party—enough, surely, to carry the party quite a bit beyond the Western frontier limitation that had strangled the Greenback and Union Labor efforts. Beyond that, the question remained: how was Populism to become the institutional voice of the "industrial millions"?

In 1886 William Lamb had implemented a boycott in support of the Knights of Labor "in order to secure their help in the near future." Obviously, the task of reaching urban workers was central to any reform movement seeking, as the Populists did, to speak in behalf of the "producing classes." Terence Powderly, the Grand Master Workman of the Knights, like his friend Charles Macune, had been swept reluctantly toward the third party. But the Knights' leader had brought few members with him. The union's membership had fallen from its peak of 700,000 in 1886 to fewer than 100,000 by the autumn of 1892. The American Federation of Labor, the new organization that had materialized from the wreckage of the labor struggles of the 1880's, was led by Samuel Gompers, a cautious strategist who focused on craft unions and avoided the more central hazards of industrial organizing. Above all, Gompers made it clear that he proposed to avoid distracting political adventures. Under the circumstances, the A. F. of L. chieftain seemed to offer even less assistance than Powderly did. To Populists, the most promising labor spokesman was a railroad union man named Eugene Debs. But while Debs had caught the attention of such Populist journals as the *Nonconformist* and the *Southern Mercury,* he had not yet placed himself in position to lead masses of American workers into independent political action.

An analysis of the political consciousness of the working class at the moment of the Populist uprising is possible—in general terms. Among working Americans, a growing sense of their own exploitation surfaced in abundance in the Gilded Age, as demonstrated by the repeated efforts of workers to find a formula that would create an enduring trade union structure. But in the industrial sector the crucial confrontation with management—the original strike for recognition of a new union—always seemed to result in the destruction of the fledgling organization that attempted the strike or, at best, in its fatal weakening. Corporate America simply possessed far too much economic and political influence, the latter manifested through friendly judges who issued timely court injunctions and, when need be, by friendly governors who could be counted upon to call out the National Guard. An innovative new instrument of social control also appeared in these years in the form of the Pinkerton organization, which, in one of the more aggressive assertions of the free enterprise spirit, sold strikebreaking for a fee. The Pinkertons were, as the Populist platform of 1892 described them, a "hireling standing army." In the late nineteenth century and on into the twentieth, the courts, the press, the National Guard, governors, legislatures, and the Pinkertons all worked in harmony to defeat workers at the pivotal moment of the recognition strike. Working Americans in the cities thus were unable to complete that necessary first step toward democratic autonomy, the creation of enduring mass institutions of their own. It was an organic problem the labor movement was not to begin to solve until the 1930's with the advent of the sit-down strike and the successful construction, by the C. I. O., of at least the beginnings of a mass base among working Americans.

At the moment the People's Party appeared, the urban American labor movement, while increasingly "aware" in economic terms, had developed no means of spreading a corresponding political consciousness to the huge working class ghettos of the nation's cities. Catholic workers tended to vote Democratic—in step with whatever ethnic leaders they had, who led whatever organizations they may have been able to construct. Among the great mass of working class Americans, thoughtful analyses of the nation's party system were, therefore, in short supply. Other than those labor leaders with Populistic inclinations, such analyses of the corporate domination of the two-party system as existed emanated from the relatively small (and in large part ethnically based) socialist faction of the labor movement. Manifestly, socialists had their own cultural problems in reaching out to the labor rank and file, not least of which was the unrelenting opposition to their efforts by the Catholic Church.

Thus, in a fundamental cultural sense, the American labor movement was simply not yet ready for mass insurgent politics. As of 1892 it had not developed, through its own institutions, a working class structure that combined economic and political consciousness in a way essential to the maintenance of an insurgent posture in the presence of the continuing cultural influences of the corporate state. Some of the more alienated partisans in labor's ranks nursed notions of the transforming educational value of "the general strike" as a means of politicizing the American working class. But such desperate dreams only testified to how isolated these theorists were, and how utterly they had failed to build an authentic infrastructure of organization among the impoverished of the nation's cities. "Organized labor" was still unorganized; it was therefore not in a position to bring masses of recruits to the new party of the people.

To hopeful Populists, of course, the possibility existed that urban workers could be reached outside of labor organizations. Edward Bellamy's humanistic novel, *Looking Backward,* published in 1887, had been widely read in reform circles and had led to the creation of a new political group called the Nationalists. The organ of the Nationalists, *The New Nation,* had enthusiastically hailed the formation of the People's Party, as had Bellamy himself, but, as of 1892, the Nationalists could rally little more than scattered groups of middle class liberals across the nation. Unorganized, but with intense feeling, the Nationalists disappeared into the mass of the People's Party. To the third party effort they added a few campaign contributions, a good deal of advice, and not much else.

However one looked at urban America—as masses of incipiently class conscious workers or as democratic citizens who could be politically activated through such appeals as Edward Bellamy's Nationalism—it is clear that the agrarian organizers who had created Alliance-Populism had not, by 1892, thought through the dimensions of the task they faced. At its most elemental political level, that task turned on the need to create among urban workers a culture of cooperation, economic analysis and a sense of autonomy that matched the political achievement of the Alliance cooperative crusade. Though Populist monetary theorists could argue that the doctrines of greenbackism would benefit urban workers, as, indeed, would all the other third party goals aimed at coping with "concentrated capital," neither the Alliance nor anyone else had ever been able to create an ongoing environment of economic education that extended to the masses of workers. Labor greenbackers represented a tiny minority of the nation's workers who, perforce, voted Democratic or Republican largely in response to old habits and old slogans.

Reduced to its essentials, the organizing problem facing the People's Party in its maiden campaign of 1892 grew out of the cultural limitations of the Alliance movement itself. What could a Protestant, Anglo-Saxon Alliance organizer say to the largely Catholic, largely immigrant urban working classes of the North? The lessons learned in the cooperative crusade simply did not supply an answer.

In 1892, what Alliance-Populism lacked was a social theory of sufficient breadth to appeal to all those who had not received an education in the Alliance cooperative. The most creative theoretician produced by the agrarian revolt was Charles Macune: he alone had combined economic analysis with the needs of organizational expansion; he alone had achieved not one, but two creative breakthroughs: the large-scale credit cooperative and the land, loan, and monetary system known as the sub-treasury plan. What the cause of the third party urgently needed in 1892 was for Macune or someone else to go beyond the social conceptions embedded in the sub-treasury system to develop a broader theoretical analysis that could be shaped to speak with power to the millions of the "plain people" in the nation's cities. Admittedly, this assignment constituted a cultural challenge of enormous dimension; indeed, it was a challenge that was to confront—and in many ways defeat—succeeding generations of democratic theorists down to the present. No more than Macune did latter-day reformers possess a language of politics that could persuasively describe to most Americans the realities of power and privilege inherent in the society they lived in.

In 1892, the prospects of urban Populism could be summarized in one sentence: the Alliance organizers looked at urban workers and simply did not know what to say to them—other than to repeat the language of the Omaha Platform. While that document could be quite persuasive in intellectual terms, few Americans understood better than veteran Alliance lecturers that organizers could not create mass institutions of reform by winning, one at a time, intellectual debates with individual citizens over the fine points of a political platform. In order for great numbers of hard-pressed people to achieve the self-confidence, self-respect, and psychological autonomy essential to a movement aiming at significant changes in the culture of a society, something more than the Omaha Platform was needed.

At root, what the People's Party lacked in the nation's cities in 1892 was the essence of Populism itself—the "movement culture" that the Alliance had created in over 40,000 suballiances across rural America in the course of building its structure of economic cooperation. This culture was, in the most fundamental meaning of the word, "ideological": it encouraged individuals to have significant aspirations in their own lives, it generated a plan of purpose and a method of mass recruitment, it created its own symbols of politics and democracy in place of inherited hierarchical symbols, and it armed its participants against being intimidated by the corporate culture. The vision and hope embedded in the cooperative crusade held the agrarian ranks together while these things took place and created the autonomous political outlook that was Populism. The labor movement, losing its recognition strikes, was simply unable to develop an organization capable of generating such possibilities of mass political cohesion. In the cities of America, therefore, the People's Party of 1892 was an institution still searching for a way to reach and talk to the people.

2

Thus, outside the ranks of the Farmers Alliance, the People's Party possessed few institutional means to attract the "industrial millions." This fact led to a vital—though apparently unconscious—shift in the tactics of the organizers who had generated the agrarian revolt in America. In effect, the Alliance founders gave away their cultural claims within the new party they had worked so hard to create.

This alteration within the reform movement began in 1892 and proceeded essentially from the successive stages of the Alliance organizing process itself. On the eve of the great lecturing campaign that carried the cooperative movement to much of the nation in 1890–92, the Alliance had been centered in eleven Southern states plus Kentucky and the Midwestern states of Kansas, Missouri, and the Dakotas. In most of these sixteen states the cooperative program of the Alliance had organized the bulk of the farmers. But the twenty-odd states added to the National Alliance after 1889 still had relatively small memberships in 1892. When the agrarian revolt moved into its political phase at Omaha, the Alliance spokesmen in the new states generally represented small pressure groups rather than mass insurgent constituencies. The organizational weakness surrounding such spokesmen presaged a kind of brokerage approach to the political process. Unable to elect candidates on their own, they tended to be drawn toward accommodation with one or the other major parties. At such moments, of course, the Omaha Platform of the People's Party constituted a hindrance, as it contained radical planks offensive to both major parties. In consequence, organizational fragility often led to a brand of politics unrelated to the greenback doctrines bequeathed to the third party by the Farmers Alliance. Indeed, brokerage politics contrasted sharply with the long-term objectives of radical Alliance organizers bent on restructuring both the American party system and the American banking system.

The immediate result of these dynamics was the unusual makeup of the party's national committee, where the men who had created Alliance-Populism were clearly outnumbered by aspiring Populist political brokers. Even the party's new national chairman, Herman Taubeneck of Illinois, came from a state containing a relatively small agrarian movement. Though the Farmers Mutual Benefit Association had achieved some organizing success in southern Illinois, the possibility of a genuine statewide agrarian presence had been crippled by the unsuccessful policies of the Northwestern Alliance. Taubeneck himself was therefore an incipient political broker. In the many subtle ways that an inherited party system affects the options perceived by individuals seeking to alter those very inheritances, the culture of America stood as an immense barrier to Populist intentions.

Interestingly enough, there is no record that the Alliance organizers voiced any objection to the leadership arrangements in their new third party. Most of the incoming representatives seemed thoroughly at home with the purposes of Populism. Were they not Populists all, whether politicians or organizers? Jerry Simpson was a politician, for example, and no stouter defender of Alliance principles could be found. So matters seemed in 1892.

In the midst of these frantic months of party-building, one other component of the reform movement was not focused upon. In ways that very few of the participants fully grasped, the fate of the individual third parties across the nation in 1892 revealed the intimate relationship between the Alliance cooperative movement and the People's Party. Where the Alliance organizers had performed their task in great depth, the third party appeared with an immediate mass base of support. On the other hand, where the organizers had reached the farmers but had failed to follow through with the organized lecturing campaign that constituted the "politics of the sub-treasury" in 1891–92, great numbers of potential Populists failed to cross the "Alliance bridge" from their old political allegiance to the new party of reform. Finally, where the Alliance lecturers, for whatever reason, had failed to recruit the bulk of a state's farmers, the third party movement in such areas was fatally undermined from the outset.*

Relative degrees of agricultural poverty did not play a decisive role in this process.† Rather, how state third parties fared depended on the extent of mass political consciousness among farmers, an ingredient that turned on how well or poorly the "politics of the sub-treasury" was conveyed through the state

* For a summary of the theoretical achievements and shortcomings of Populist organizers see pp. 297–310 and pp. 318–19.

† For example, farmers in Kansas, Nebraska, and Texas, working newer and less exhausted land, were marginally "better off" than farmers in South Carolina, Georgia, Alabama, and Mississippi. Yet Texas, Kansas, Georgia, and Alabama produced highly visible Populist insurgencies, while Nebraska, South Carolina, and Mississippi did not. Mass protest movements turn on organizational components governing the presence or absence of mass political consciousness, not on mere gradations of "poverty." The latter is only secondarily relevant. Interestingly, while economic determinism is an outgrowth of Marxist thought, its more sweeping interpretive uses have been the special province of capitalist historians, i.e. Populism began with "hard times" and ended when "prosperity returned." (For the non-existence of agrarian "prosperity" after Populism, see pp. 268–69 and pp. 297–98.)

On the other hand, the ideology of the third party did not turn on the "class" composition of individual state Alliances. A higher ratio of "landless" to "landed" farmers in a given state Alliance did not organically determine the extent to which a Populist presence materialized in that state. The determining factor remained political consciousness of individual farmers as to the unserviceability of the inherited two-party system. Where the sub-treasury was not made the test of Alliance loyalty, as in Tennessee, huge state Alliance organizations, with adequate numbers of landless and landed farmers, produced only a fractional Populist presence. In Alabama, where Populism was relatively strong, and in Tennessee, where it was weak, the point of unity was the fact that the Alliance enrolled the overwhelming majority of *all* farmers, "landed" and "landless" alike. The dramatic difference in the strength of Populist protest in these adjoining states was not organically related to gradations of class. The movement culture of shared experience within the Alliance imparted special credibility to sub-treasury lecturers, making mass insurgency possible. But if this culture were not capitalized upon—if the lecturing campaign on the sub-treasury did not take place—mass insurgency did not result, whatever the facts of class or poverty in a given state. There were no exceptions in the states of the South and West. The subject is further discussed on pp. 139–44, 164–65, 215–22, and 304–5.

alliance lecturing system. Thus, while the cooperative movement was always central, the political posture of individual state leaders had much to do with the kind of Populist presence that materialized. The fate of the People's Party across the nation in 1892 turned on these relationships: the political movement was the movement politicized—nothing more, nothing less.

In the South, the politics of the sub-treasury had been brought home with varying degrees of organizational thoroughness and intellectual coherence to the farmers of Georgia, Alabama and Texas; less so in North Carolina; and not at all in Tennessee and Florida. Ben Tillman succeeded in destroying the South Carolina Alliance in 1892, while the order's lecturers had never built a statewide mass movement in Virginia and Louisiana. In all these states, the third party cause was heavily burdened in 1892 with a national ticket headed by a former Union general. Sectional loyalties, therefore, concealed the depth of the Alliance organizing achievement; in 1894, with no national ticket, state third parties with strong cooperative bases in the South grew surprisingly.

The same dynamics described the third party's fate in the West. The relative thoroughness with which the doctrines of greenbackism had been absorbed by thousands of farmers in Kansas and the Dakotas ensured a Populistic dedication to the Omaha Platform, even in those instances where temporary alliances with Democrats were affected. But in Nebraska, the absence of the cooperative experience left a political movement wholly unrelated—and unresponsive—to the greenback doctrines of the Omaha Platform. The extremely shallow base for the third party in Illinois, Wisconsin, Iowa, and Minnesota left by the Northwestern Alliance had only been partially augmented by the belated organizing campaign of the National Alliance in 1891–92. Similarly, Alliance organizers had raised the banner of cooperation from the mountain states to California in the national lecturing campaigns of 1890–92. Where the shoots had taken root—especially in parts of Washington, Oregon, California, and eastern Colorado—impressive numbers of farmers became Populists. In mining states, the agrarian base was augmented by miners who liked the free silver plank of the

Omaha Platform. In such instances, of course, the miners were no more imbued with the greenback principles underlying the cooperative crusade than their agrarian neighbors in Nebraska were. Throughout both the South and the West, therefore, the strength and sense of purpose of Populism was directly related to the strength and sense of purpose of the cooperative crusade that had created the reform movement.

In the heartland of the Western granary, along the farming frontier from Dakota to Kansas, the new third party had a complicated inauguration. The shadow movement of Populism in Nebraska immersed itself in Democratic fusion—and on Democratic terms. A modicum of success resulted. The fusionists re-elected Omar Kem and William Jennings Bryan to Congress, giving each party a man of its choice, and supported a third "straight-out" fusionist who won. The Republicans swept the state offices from the governorship on down and won the other three congressional seats and the state's electoral votes for Harrison. The ideological disarray in these proceedings caused surprisingly few internal tremors—a clear indication that the third party in Nebraska, having failed to generate a culture of reform, possessed few reform principles it considered important enough to defend.

Despite their remarkable showing in 1890, the Kansas Populists entered the 1892 campaign with a certain anxiety. Local municipal elections across the state in 1891 indicated that the "People's" ticket was more difficult to put across to voters than had been the "Alliance" ticket of 1890. The setback caused a shift in policy. Officeholders, who were rather prominent in the Kansas third party leadership, responded to the election returns by sounding out Democrats on the idea of possible cooperation. For the governorship the Populists nominated L. D. Lewelling, a Wichita merchant and ardent egalitarian who had advised the nominating convention to "welcome honorable allies" and "go forth to victory." Advocates of fusion, though heavily outnumbered in the Kansas People's Party, were hard at work in the state. Except for two congressional districts where bona fide three-cornered races were in prospect, the Republicans faced a united Populist-Democracy running on the basis of the Omaha

Platform. The third party's adroit, pragmatic state chairman, John Breidenthal, soon converted both races into two-sided affairs by persuading the Democrat to withdraw in one district and the Populist to do so in the other.

The resulting victory of the People's Party was, or appeared to be, something of a sweep. Lewelling and the entire Populist state ticket won, though by a narrow margin of 5000 votes out of 320,000 cast. The People's Party had gained control of the state senate, but on the face of the Republican-certified returns, the Populists had unaccountably lost their huge margin in the House. A comparison with the 1890 returns reveals that the total of Populist and Democratic votes had declined, from 178,000 to 163,000, while the Republican total had climbed dramatically, from 115,000 to 158,000. Cooperation with Democrats on the basis of the Omaha Platform had staved off defeat, but the great flood of votes in 1890 that seemed to presage a new kind of democratic politics emanating from the American heartland had not been augmented. The rest of the nation interpreted the election as a Populist triumph, but the "party of plutocracy" was able to poll within 1 per cent of its combined opposition. Such resiliency was sobering. Sectional loyalty to the old party was remarkably enduring. Was the "coming revolution" but a fragile illusion?

But this perhaps was an overly pessimistic view. The Greenback and Union Labor parties had never achieved the kind of organic presence across the region that the new People's Party attained in 1892. Even in the Dakotas, where Republican partisans frantically waved the "bloody shirt" in an effort to keep the citizenry from focusing too closely on economic issues, the People's Party emerged in a far stronger position than had any previous reform institution. The new ideas of the Omaha Platform had clearly reached many voters.

Evidence appeared in Western states all the way to the Pacific coast. In the mining states, Populists sometimes nominated—and, in Colorado, elected—dedicated greenbackers. But the rank and file of the miners were responsive to another and narrower plank in the Populist platform—the call for the free coinage of silver. In Colorado a radical Aspen editor named

Davis H. Waite was nominated for the governorship by the Populists, endorsed by the Democrats, and elected in November. After an 1893 speech in which he said "It is better, infinitely better, that blood should flow to the horses' bridles rather than our national liberties should be destroyed," he became widely known as "Bloody Bridles" Waite. Plagued with a Republican-controlled House, Waite could achieve little in the way of lasting reform. The state legislature not only refused to pass the strict railroad regulation bill he proposed, but also repealed the existing law. Waite's subsequent conduct in a complex and bitter strike in the gold-mining district near Cripple Creek averted probable bloodshed, but won him no plaudits from Republicans. In any event, though Waite's broadly democratic ideas have been called "those of a fanatic," ideas as such were not the organizing element in Populism in Colorado. The emotional pull of "free silver" was.

The same single-mindedness characterized the third party in Nevada where, in fact, it was known as the "Silver Party." The victorious Silverites nevertheless gave their three votes in the electoral college to Weaver. In Idaho the free-silver Populists were less successful as a state party, but hardly less so in the presidential contest. The triumphant silver Democrats bypassed Grover Cleveland and cast all their electoral votes for Weaver.

The third party had a broader thrust in Montana, where it primarily represented a movement of the urban working class, augmented by farmers mobilized by the newly arrived National Alliance. These groups formed the backbone of the Montana People's Party, which ran a fairly strong third to the two major parties in 1892. The strongest Populist vote-getter in Montana was the candidate for Attorney General, a woman's rights advocate named Ella Knowles.

Farmer-labor coalitions were also achieved in the Pacific Northwest, where the agrarian element played a larger role than it did in the mining states. Polk's organizers had arrived in Washington in July of 1890 and had established some 200 suballiances by the following year. An Alliance Implement Company was capitalized at $100,000 and a state journal, the *Alliance Manifesto*, inaugurated at Spokane. In 1892, the third party nominated a Whitman County farmer who ran particularly well in the Alliance strongholds of eastern Washington. The victorious Republican received 33,000 votes to 29,000 for the Democratic and 23,000 for the Populist aspirant. The agrarian-based third party's maiden performance in state politics both surprised and impressed Washingtonians.

The Alliance, with the oratorical assistance of "Cyclone" Davis, arrived in Oregon in 1890 and came under the aggressive influence of what proved to be one of Populism's hardest working couples, Seth and Sophronia Lewelling. The Oregon People's Party was formed in the spring of 1892, on the day following the state Alliance convention. The new party, strongest in the farming districts, ran third in 1892, as did a somewhat out-manned but militant third party in California. Following the election, the California party immediately began putting out feelers to working class elements in San Francisco and other cities as a step toward a workable farmer-labor coalition. Signs of cooperation appeared that augured well for the third party's future.

Nevertheless, not all the indicators in the West were positive. Some of the biggest disappointments of the 1892 campaign came in what had been counted on as potentially fruitful third party territory in the Old Northwest. The People's Party did unexpectedly poorly in the entire Great Lakes region extending through Minnesota, Wisconsin, Michigan, Iowa, Illinois, Indiana, and Ohio. In those states, which comprised much of the original territory of the Northwestern Alliance, what little agrarian strength that materialized in 1892 largely stemmed from the efforts of the National Farmers Alliance. This organizing work, while encountering very little counter-organizing by the Northwestern group, did meet with a fair volume of rhetorical resistance. The Republican secretary of the moribund Wisconsin affiliate of the Northwestern group issued a frantic announcement in 1892 that employed the language of bloody shirt politics. Addressed to the "loyal Alliances of Wisconsin," the missive deplored the latest defection from the ranks—that of Ignatius Donnelly—characterizing the event as a "sell-out" to the Southern-based order. While such strictures could have small effect

on the Wisconsin membership—it being virtually nonexistent—
the latent sectional issue produced confusion, and confused
farmers proved hard to organize. The People's Party thus had
to build upon a very shallow base in much of the Old Northwest
in 1892, and the election returns indicated that very little work
had been completed prior to the November elections.

Even in Minnesota, where matters were under the management of Ignatius Donnelly, the author of the ringing preamble
to the Omaha Platform, Populism failed to capture the expected
following. Engaged in constant internal wrangling within his
faction-ridden Minnesota Alliance, Populism's foremost novelist
learned in 1892 the limits of rhetorical display as an instrument
of social change. Despite an exhausting speaking campaign, he
led the People's Party to a convincing defeat. In the presidential
race, Minnesota gave Weaver 11 per cent of its vote. Utterly
disconsolate, Donnelly wrote in his diary: "Beaten! Whipped!
Smashed! . . . Our followers scattered like dew before the rising
sun."

Elsewhere the third party showing was even more feeble.
Weaver received less than 5 per cent of the vote from Michigan
across to Ohio. Most humiliating of all, he did equally poorly in
his native state of Iowa. The Old Northwest still had plenty of
agrarians, but no hope of an agrarian movement. The organizing
prerequisites that were essential if the heritage of sectionalism
were to be overcome had simply never been fulfilled.

4

But whatever happened in the upper Midwest in 1892, Populism's greatest test, as all radicals understood, lay in the American
South. The idea of a great multi-sectional party of "the industrial
millions" had always shattered on the rock of the solid Democratic
South. For an answer, radical strategists did not have long to
wait; the new party's very first test came a month after the
Omaha convention—in the Alabama state elections of 1892. The
contest revealed the stresses and strains awaiting Populist candidates across the South during the life of the agrarian revolt.
The state's chief agrarian spokesman, a former state agricultural

commissioner named Reuben Kolb, moved cautiously into insurgency. In the time available to it in 1891–92, the large Alabama Alliance tried manfully to employ the sub-treasury issue as the driving wedge. But one Alabama writer reported that "many Alliance members did not understand the sub-treasury plan." Nevertheless, the Macune plan found many willing advocates in the state, and the sub-treasury message was brought home, if not to all farmers, certainly to many.

Kolb was chosen as the nominee of the "Jeffersonian Democrats" at a statewide convention dominated by the Alliance, and incumbent Governor Thomas Jones was renominated by the conservatives. The resulting election was as fraudulent as the nominating process had been earlier in the year. On the face of the returns, Jones received 126,949 votes to Kolb's 114,424. But, as an unusually candid Democratic paper conceded, "the truth is that Kolb carried the state, but was swindled out of his victory by the Jones faction, which had control of the election machinery and used it with unblushing trickery and corruption." It developed that a largely fictitious black vote had been decisive. In the "black belt" counties, which ranged up to 80 per cent Negro in population but had long been dominated by the planter faction, huge majorities for Jones offset Kolb's margins in the poor-land counties, where white farmers constituted the majority. Twelve black belt counties returned majorities of 26,000 for Jones— more than double his statewide margin. As a later congressional investigation showed, "Negroes who had been dead for years and others who had long since left the county" somehow voted Democratic. In the upland counties, many of which had seen the growth of an active Colored Farmers Alliance and a period of cooperation between farmers of both races, the majority of black farmers voted for Kolb along with their white compatriots. Indeed, even in one black belt county, Choctaw, the Colored Farmers Alliance was well enough organized that a majority of blacks voted for Kolb despite all the outside pressures that were brought to bear.

In Georgia the third party sought to lead with its strongest candidate by awarding its gubernatorial nomination to Thomas Watson. But when Watson decided to run for re-election in his tenth congressional district instead, the Georgia Alliance turned to the guiding spirit of its statewide cooperative exchange, "a real dirt farmer" named W. L. Peek. A man of considerable ability, Peek was not at his best on the campaign trail, but he did symbolize the commitment of Georgia Alliancemen to their order, to the cooperative crusade, and to the sub-treasury plan. The final defeat of the views of Lon Livingston inside the Georgia Alliance had been clear since March 1892, by which time no less than 1600 of the 2200 Georgia suballiances had explicitly endorsed the full Ocala Platform in an impressive display of farmer solidarity. Nevertheless, in the fall elections national attention focused not on the governor's race, but on Tom Watson's tenth district. The pro-Alliance Congressman's outspoken attacks on the two old parties in Washington had earned him national attention—and symbolic status as a barometer of the third party's future in the deep South. Watson's Democratic opponent, Major James Black, invoked the familiar appeals of Southern sectionalism, stressing his own Civil War service and the need for white Southerners to stay with the old party to avoid the possibility of Negro "domination." The Democratic press, solidly behind him, raised the specter of Reconstruction and the "revival of bayonet rule." Watson's ability to overcome such Democratic tactics seemed so symbolically linked with the Southern future of the third party that though the ensuing campaign was waged in a single congressional district no less a personage than Grover Cleveland confided that "he was almost as much interested in Major Black's campaign in the Tenth District of Georgia as he was in his own election." Businessmen in Augusta, the district's principal city, made special appeals to their financial connections in New York City on the grounds that Watson was "a sworn enemy of capital, and that his defeat was a matter of importance to every investor in the country." The *New-York Tribune* reported that railroad and insurance interests responded "liberally" to this appeal, augmenting local Georgia money with $40,000.

The intensity of the campaign was unparalleled. The Atlanta *Constitution* felt that the threat of "anarchy and communism" extended to the entire South because of "the direful teachings of Thomas E. Watson." At one point Watson's supporters rode all night to rally to a black Populist who had been threatened with lynching; the Augusta *Chronicle* was too outraged to feel the need for a bill of particulars. "Watson has gone mad," the paper decided. The conduct of Democratic managers on election day was so openly fraudulent that one historian specified "intimidation, bribery, ballot-box stuffing, and manipulation of the count," as applicable adjectives, while another settled for "terror, fraud, corruption, and trickery." In Major Black's sole bastion of support, Richmond County, containing the city of Augusta, Democratic majorities soared to 80 per cent. The total vote solemnly returned by election judges exceeded double the number of legal voters. On the face of the returns, Richmond County provided Black with enough votes to offset heavy Watson majorities in the surrounding rural counties and provide the margin for a Democratic "victory."

Whether the white vote had been counted fairly or not, it was clear to all that blacks now held the balance of power in Georgia. A number of murders occurred—no one knows how many. At Dalton, a Negro man who had spoken for Populism was killed in his home, and a black minister who repeatedly spoke for Watson was fired upon at a rally, the errant bullet striking a near-by white man and killing him. Election day murders took place elsewhere, particularly in Watson's tenth district. Only in such ways did the Democratic Party hold its lines intact in Georgia in 1892.

In Alabama and Georgia, men like Kolb and Watson contested orthodox leaders of the conservative "New South." In Texas in 1892, the third party faced a more complicated political opposition. The Texas Democracy was split into two factions—a dominant "reform" wing, largely based on Alliance voters and headed by Governor Hogg, and a "goldbug" faction, led by railroad lawyer George Clark. While the Democratic division seemed to augur well for the third party, Hogg's platform style and his apparent dissociation from the railroad wing of the party cut heavily into the Alliance vote.

The Republican Party, which numbered some 75,000 to 90,000 black voters among its constituency, played a decisive but enormously complicated role in the 1892 campaign. The three contending parties competed fiercely for the black vote, the Populists through the structure of the Colored Farmers Alliance and the newly forming Populist clubs across the state, the "Railroad Democrats" through a "top-down" arrangement with Negro Republican leader Norris Wright Cuney, and the Hoggites through the time-tested Democratic methods of bribery, intimidation, and overt violence.

The third party nominated as its gubernatorial standard-bearer Judge Thomas Nugent. He was easily one of Populism's most striking contributions to American politics. A soft-spoken man, Nugent possessed qualifications not normally seen on Southern political hustings. Widely read in both economics and religion, his speeches were studded with quotations from Immanuel Kant, William James, John Stuart Mill, David Ricardo, and Herbert Spencer, in addition to such Populist favorites as Thomas Jefferson, Thomas Paine, and Edward Bellamy. Despite a deep, personal radicalism, Nugent articulated the Populist creed without rhetorical flourish. He addressed himself to the "labor question" and the "problem of the distribution of wealth," which he considered paramount among the nation's political issues, but he also stressed the socio-theological values he felt the Republic and its citizens needed to achieve. A Swedenborgian in religion, Nugent was essentially a radical humanist. His bald head, flowing beard, and quiet but confident manner gave him the aura of a prophet.

But in Texas, as elsewhere in the nation, ridicule had an irresistible appeal to politicians who preferred not to discuss the causes of poverty on Populist terms. Nugent calmly turned aside such thrusts by accepting the charge of unorthodoxy—and then linking unorthodox ideas to the teachings of Jesus, one who "did not hesitate to denounce wrong, even though [it was] hedged about and protected by social power and influence." In Nugent's hands, Populism was inextricably linked both to socialism and to a carefully defined Prince of Peace: "Jesus saw the fatal tendency of men to think in customary and institutional lines. . . . Here was the beginning of Christian socialism . . . vital fundamental truths thrown down upon the current of public thought and sent drifting down the ages. Was Christ the consummate product of divine evolution led forth into the human world to transform and uplift and glorify the social Man?"

Texans were not used to hearing their politicians conclude lofty passages with question marks; it took a while for the "Nugent tradition," as it came to be known, to catch on. The three-way 1892 campaign saw the contenders divide the white vote fairly evenly, but both Clark and Nugent were snowed under by black votes for Hogg. The returns gave Hogg 190,000, Clark 133,000, and Nugent 108,000. In the score of Texas counties containing heavy concentrations of black voters, Nugent ran a poor third, sometimes receiving less than 1 per cent of the vote. Intimidation and terrorism proved as effective in Texas as they had in Alabama and Georgia.

If something on the order of half the Alliance membership in Texas had opted for the People's Party at the first opportunity, reformers there seemed merely to increase their efforts to win the remainder in the off-season. Their achievements provide an interesting footnote to the agrarian revolt and to the history of democratic reform movements in America. Veteran Alliance lecturers, augmented by a steadily expanding reform press, carried the radical doctrines of the Omaha Platform to the farthest reaches of the state. Elaborate summer encampments— a Populist folkway that hugely discomfited Democrats in Texas— began to reach alarming proportions in 1893–94 and the growth of the ranks of reform became obvious to everyone in Texas. In the spring of 1894 the Hoggites lost their poise. The Texas Democracy hurriedly closed ranks, the "reform" forces of Jim Hogg making sweeping concessions to the railroad wing of the party in exchange for party unity.

In the 1894 elections the Texas People's Party set the stars of the Alliance lecturing system upon the party of the fathers. Evan Jones, J. M. Perdue, and "Cyclone" Davis, along with Jerome Kearby, a gifted Dallas lawyer who had defended the Knights of Labor leaders in the Great Southwest Strike of 1886, ran for Congress. William Lamb's Reform Press Association of 1891 mobilized over a hundred editors by 1894, and "Stump" Ashby, the state party's chairman, presided over a corps of orators that included John B. Rayner, a huge, three-hundred-pound black apostle of Populism, and Melvin Wade, a radical black trade unionist. While Wade worked in the cities, Rayner, a tireless organizer, trained scores of local lieutenants to carry the doctrines of reform into the rural underside of the Republican Party apparatus even as Democratic leaders sought to make financial and patronage bargains at the top with black Republican managers. The third party more or less wrote off congressional districts in the southern one-fourth of the state, as the instruments of social control in the Mexican-American districts simply proved too formidable for the reformers to overcome. On election day, the third party carried vast sections of North and West Texas and dominated in parts of the central and eastern regions of the state—seven congressional districts in all. The Populists were counted out in every one of them. The methods were the same as in Georgia and Alabama—wholesale ballot-box stuffing, open bribery, various forms of intimidation, and massive voting by dead or fictitious Negroes. The Richmond County methods of Georgia were almost precisely duplicated in the "Harrison County methods" used in East Texas to defeat "Cyclone" Davis. Indeed, in Texas the phrase "Harrison County methods" became the standard term defining the most effective Democratic campaign technique of the Populist era. Even on the face of the returns, and including in the total the controlled vote of South Texas, the Populist vote jumped from the 23 per cent of 1892 to almost 40 per cent in 1894. The "official" statewide total showed Nugent had been defeated for the governorship by 230,000 to 160,000, though a number of steps were taken to ensure that the real outcome would be forever beyond recovery. The Populist candidates went to court—unsuccessfully—and the *Southern Mercury* led a chorus of indignation in the Texas reform press. Meanwhile, the third party faithful shrugged their shoulders and renewed their off-year "educational" campaign. The trains of farm wagons grew ever longer, and the Populist encampments of 1895 were the largest the state had ever seen.

Nowhere else in the South, however, was the Alliance able

immediately to transfer its constituency to the third party on the scale achieved in Georgia, Alabama, and Texas. The reasons varied, and they sometimes seemed to stem from contradictory causes. The fate of Populism in Louisiana, however, was scarcely puzzling; indeed, given the interrelated character of that state's fraud and violence, the story of Populism there is most coherently told as a single piece. The rather slow Populist entry in Louisiana politics in 1892 was followed by unmistakable signs of growth in 1894 and then by an impassioned struggle in 1896 that ended in anarchy by conservatives and a threat of civil insurrection. The fledgling People's Party had to compensate for an incomplete Alliance organizing effort in the 1880's that was at least partly traceable to the lack of skill of its founding leader, J. A. Tetts. The agrarian cause was further hampered in 1892 by byzantine factional maneuvering growing out of a campaign to end Louisiana's infamous state lottery. In all the confusion such issues as the Omaha Platform and the sub-treasury plan never gained much clarity in Louisiana, despite zealous efforts by the leader of Louisiana Populism, a minister's son from Winn Parish named Hardy Brian. He antedated a somewhat better known agrarian whom Winn Parish would offer the nation in the twentieth century—Huey Long. The storied career of the Kingfish, in fact, is rendered partly understandable by the prior experiences of Brian in the Populist era.

Well-organized Populist congressional campaigns in two hill-country districts in 1894 were defeated through varieties of ballot-box thievery that in sheer arrogance and venality surpassed similar depredations in the rest of the South. Beginning with its founding meeting, which brought together members of both the white and the black Alliances, the Louisiana third party had developed a most un-Southern proclivity for interracial cooperation; after the 1894 campaign, alarmed Democrats concocted an elaborate "election reform" law to disfranchise thousands of impoverished farmers of both races. Populist resistance to this proposal, led by Brian, was of a character that inevitably prepared the way for a Populist-Republican coalition in 1895–96. This combination, though somewhat vague and sometimes even contradictory in the economic principles enunciated by its variegated spokesmen, posed a clear threat to the state's ruling oligarchy for the first time since Reconstruction. The conservative press promptly christened it the "Populist-negro social equality ticket." The Shreveport *Evening Judge* announced that "it is the religious duty of Democrats to rob Populists and Republicans of their votes whenever and wherever the opportunity presents itself and any failure to do so will be a violation of true Louisiana Democratic teaching. The Populists and Republicans are our legitimate political prey. Rob them! You bet! What are we here for?"

The gubernatorial candidate of the coalition, John Pharr, might best be described in modern terms as a liberal Republican planter. He had a habit of ignoring prevailing mores. The anti-lynching plank in his platform convinced the New Orleans *Daily States* that he "inferentially approved" of white women being raped. Similarly, the Baton Rouge *Daily Advocate* characterized the eminently respectable candidate as an "ignorant and low bred boor" who "proceeds from place to place scattering his fire-brands among the rabble and inciting the baser passions of the populace." Another Democratic newspaper tried to bring Negro political activity to a halt by printing the names of all black leaders who were "brewing up trouble," but the next issue of the paper revealed that the effort had been unsuccessful. "You might as well talk to a brick wall as to try to make the nigger believe who his best friend is," the paper complained. The "baser passions" that the Baton Rouge paper had warned against did, indeed, surface in Louisiana in 1896, though the source of these emotions seems to have been more readily fixed among the paper's subscribers than its political opponents. A multifaceted terrorism prevailed. Twenty-one lynchings occurred in Louisiana that year, one-fifth of the total for the entire nation.

Physical intimidation having failed to achieve the desired end completely, Democrats were forced to steal the governorship through massive and transparent ballot-box frauds. The victorious Democratic governor was pleased to have retained "control of affairs" in the hands of what he described as "the intelligence and virtue of the State" over "the force of brute numbers." The legislature, narrowly Democratic, certified the "official" returns, and the campaign of 1896 became history. Though the massive frauds demoralized most Populists, Hardy Brian kept up the struggle until 1899, when he published the final issue of his Populist newspaper. The following year another Winn County candidate running on a Populistic platform was defeated for the state legislature. It was to be this candidate's son, a young boy at the time, who three decades later would write the final apocalyptic paragraph to the rule of Louisiana's lawless oligarchy. Under the circumstances, it is not too much to say that Louisiana's "better elements" received from Huey Long an even milder regime than they had earned through their own actions during the earlier decades of Reconstruction and Populism.

The fate of Populism in South Carolina, Tennessee, Florida, Virginia and Arkansas revealed the variety of ways a reform movement could be sidetracked, or could sidetrack itself. In South Carolina, "Pitchfork Ben" Tillman, "the farmer's friend," destroyed Populism after a tense and demagogic campaign in 1891–92. Tillman endeavored to keep his maneuverings against the Alliance obscured by riveting public attention on his "war" with South Carolina's planter Democracy. However, his confidence in his own oratorical skill led him to a near-fatal blunder in 1891, when he accepted a debate on the sub-treasury plan with the national lecturer of the Alliance, Ben Terrell. The scale of the risk lay in the site of the debate: it was held before the South Carolina Alliance at its annual meeting. Though Tillman tried to use friends inside the Alliance to set a proper climate for the debate, the order's state president declared that South Carolina farmers could be "counted on to stand squarely by all the demands of the Alliance, Governor Tillman to the contrary, notwithstanding."

In this unusual struggle between an agrarian organization and a self-appointed agrarian spokesman, the Alliance lecturing system became crucial. On the eve of the debate county Alliances one after another endorsed the sub-treasury plan by lopsided majorities. The Tillman-Terrell debate itself took place on July 24, 1891. Tillman culminated his attack on the sub-treasury by calling it "socialistic." Alliancemen, for once, were not beguiled by their elusive Governor. They promptly endorsed the Ocala

Platform, sub-treasury and all. The pro-Bourbon Columbia *Daily Register* gloated that Tillman had been made "to eat a good, large slice of humble pie." Shaken, Tillman confided a new strategy to a close associate: "I cannot speak now without saying something which may be tortured into an attack on the Alliance and its platform, hence my determination to remain absolutely silent and let things drift."

In this immobilized position, Tillman found himself unable to lead South Carolina's farmers away from their Alliance demands. So, in the spring of 1892, he made a great show of capitulating on the sub-treasury issue in a speech to an Alliance assembly. He would stand by "all of the Alliance demands," he announced. The move was successful. South Carolina farmers rejoiced and carried Tillman from the platform. "Well, 'I went and done it,' " Tillman exulted to a friend, "My Alliance brethren are happy as the father of the prodigal son." The wily Governor followed with a *coup de grâce*—the Tillman-dominated Democratic Party endorsed the full Ocala Platform in its state convention! The state's Bourbon press was outraged, of course, but Tillman had other game in mind: the political isolation of the Farmers Alliance. By this process the third party movement in South Carolina became immobilized. After Cleveland's nomination party loyalist Tillman supported the hard money candidate of the national Democratic Party. Talking fast, "the Pitchfork man" said that he and his supporters would "eat Cleveland crow," but only after making the state's conservatives "eat Tillman crow first." And they did. Cleveland and Tillman won by overwhelming margins, while the Populist presidential ticket received only 2410 votes in the entire state. Dazed by it all, the leaders of the agrarian movement found themselves either outmaneuvered and cornered or forced to go along with Tillman—even as their organization lost its internal cohesion and political identity. In the course of the fall campaign of 1892, the huge South Carolina State Farmers Alliance, which housed one of the most impoverished constituencies in the South, collapsed. The politics of the sub-treasury had worked, but Tillman had co-opted the Alliance platform and, through effective demagoguery, had destroyed the coherence of the reform movement.

The Tennessee and Florida Alliances, on the other hand,

destroyed themselves. The 1890 Alliance successes in both states installed in office prominent Alliancemen who subsequently proved quite reluctant to risk alienating part of their support by breaking with the Democrats. In Tennessee Governor Buchanan and his successor as Alliance state president, John McDowell, vacillated so long on the sub-treasury issue that by the time McDowell committed himself to the third party, a week before the Omaha convention, it was much too late to begin a lecturing program to educate the farmers on Populist issues. McDowell was unable to win over his colleague until August 1892. After a hastily organized campaign, the embryonic Tennessee third party, with Buchanan at its head, suffered an overwhelming defeat. Florida Alliancemen, influential in the councils of the Democratic Party, persuaded old party regulars to take a leaf from Tillman's book and endorse the Ocala Platform, a development that left the order's radicals with no opportunity to employ the politics of the sub-treasury. With Alliance farmers remaining loyal to the party of the fathers, Florida Populism was stillborn.

The cooperative movement in Virginia, cautiously modeled along the Rochdale lines of the Grange, failed either to recruit or to instruct large numbers of the state's farmers about the underlying relationships between debtors, creditors, and a contracting currency. A belated push on the sub-treasury by the agrarian leadership in Virginia was not successful at the state level until October 1891, too late for the suballiances to achieve any widespread education on the issue prior to the election season of 1892. The slow pace of such necessary developments in Virginia betrayed a loosely organized state structure, a thin lecturing system, and an absence of activist thrust among the leadership. The story of Populism in the Old Dominion may be fairly summarized in a single sentence. Lacking a strong Alliance organizational base among the state's farmers, the third party simply failed to achieve a genuine statewide political presence. In the one congressional district containing strong Alliance organizations, the Populist candidate won his race in 1892, but was counted out through fraudulent returns.

In Arkansas, by contrast, the new party clearly suffered from the abrasive militancy of the early years of the state Agricultural

Wheel, when radicals bypassed thousands of "nonpartisan" Democrats. The close elections of 1888 and 1890 were revealed in 1892 to have been based on a sizable proportion of Negro support. The 1892 campaign, which involved separate Republican, Democratic, and Populist tickets, saw the People's Party run a poor third.

The North Carolina People's Party was unique in the nation. A bizarre juxtaposition of agrarian numerical strength and leadership hesitancy on the sub-treasury resulted in 1892 in huge defections from the Alliance, a development that was immediately translated into a crushing defeat for the third party in what had been regarded as one of the strongholds of the Southern Alliance. The sudden death of Polk, a decision by the Democrats to nominate the Alliance state president, Elias Carr, for the governorship, and a confused sequence of maneuvers by Marion Butler, Polk's reluctant successor as third party spokesman, all contributed to the massive split in the Alliance.

As it soon became apparent that the split was permanent, Butler boldly decided on a policy of overt Populist-Republican fusion. The tactic was implemented through a wondrous blend of "practical politics" and—occasionally—a genuine effort to build an interracial political coalition on a cornerstone of election reform to protect the freedom of the ballot. To the consternation of the Democrats, the two-party cooperation produced a Populist- Republican majority in the 1894 North Carolina legislature. When that body convened to elect two United States Senators— one for a regular term and one to fill an unexpired term—the managers of the coalition were able to hold their forces together long enough to send Populist Butler to Washington for the full term and a Republican for the unexpired term. Out of schism and desperation, North Carolina Alliancemen had buttressed their sagging organizational structure with new Republican support and had achieved, momentarily at least, a stunning victory on what appeared to be socially radical terms.

Across the Old Confederacy, the campaigns of 1892 and 1894 revealed both the reforming energy of Southern Populism and the power of the received culture. Both were shaken by the fury of the encounter. As the full impact of sectional attacks fell upon Southern Populists in 1892, the farmers of the region found the pull of the old party harder to resist than did their counterparts in the West. The absence of L. L. Polk at the head of the national ticket may have diminished the third party vote by as much as one-third to one-half in many areas. But in 1894, to the dismay of Democrats, Southern third parties—with no Union general yoked to their state tickets—grew alarmingly. The longer the third party survived, the more it weakened the political habits of the one-party heritage. By 1895 Democrats from one end of the Old Confederacy to the other had been forced to acknowledge the remarkable zeal and continued growth of the reform movement. To all who could read the signs—and Democratic Congressmen read them with growing fear—Southern Populism had become a power to be reckoned with. Though sectional politics had been employed along with steadily increasing demagoguery and terror in Georgia, Alabama, Mississippi, Louisiana, Texas, and Arkansas, the fact was nevertheless indisputable: in political terms, the "Solid South" had become a contested region.

5

As a national enterprise by a new third party, the politics of reform clearly involved great organizational and tactical complexities. To ensure voter loyalty a new party could not rely on memories, sectional or otherwise; it had to provide reasons— and the reasons had to include some kind of proof that the party had come to stay. Still, Populists were by no means disheartened. They could remind one another, and with complete accuracy, that the infant prewar Republican Party had leaped from obscurity in 1854 to national power in 1860, largely as a result of the hopelessly anachronistic character of the old Whig Party. Why should reformers feel discouraged? In 1892 *both* major parties appeared anachronistic! Both were in thrall to the whims of the money power, and "concentrated capital." The "Gold Democrats" had narrowly defeated the "Gold Republicans" and given the nation President Grover Cleveland. Both old parties were continuing to turn their backs on economic realities and were working in a harmonious "sound money" partnership to "down the people." Despite all the hazards of a maiden campaign, third party leaders found reason for optimism. Their presidential candidate, General James Weaver, had received over one million votes, and five states had cast one or more electoral votes for the Populist nominee, giving him a total of twenty-two in the electoral college. Weaver himself publicly characterized the showing as "a surprising success." Moreover, the thin Western base of the Greenback Party years had been vastly expanded; in parts of the South the voice of protest was remarkably loud, even when temporarily muted by illegal means. Clearly, the arrival of the People's Party had come with sufficient force to attract the attention of political observers throughout the nation.

Yet the sectional barrier to independent political action had shaken the organization of the National Farmers Alliance at every level—including, it turned out, its highest. Indeed, immediately following the November election it became starkly apparent that the new party desperately needed whatever stiffening it could get from Alliance organizers and reform editors—for the final days of the campaign had uncovered an unanticipated organizational crisis of the first magnitude.

6

If the third party had drawn its most promising strength from those states in which the National Alliance had enlisted the most farmers, it was also demonstrably clear in November 1892 that the one man who had done so much to build that membership, the editor of the order's national journal and its founding president, Charles Macune, had slipped out of the fold: in the final days before the election Macune had decided he was not a Populist after all! At the order's national convention—held in Memphis shortly following the November election—the extraordinary details of Macune's defection exploded to the surface.

It turned out that the *Economist* editor's reluctant conversion to the third party cause at the Omaha convention in July had not survived the early fall state election in Georgia. To Macune that election seemed to indicate that "the party of the fathers" was destined to hold the loyalty of most of its Alliancemen in the South. Macune apparently saw this development as a fatal threat to his power base in the Southern Alliances and a possible curtailment of his prominent—and salaried—role in the agrarian movement. Late in October, just before the national election, Macune acquiesced in the shipment of pro-Democratic campaign literature from Alliance national headquarters in Washington. The mass mailing, which, predictably, created a furor among third party men throughout the nation, bore the signature of J. F. Tillman, the Alliance national secretary, who was a loyal Macune functionary. At the Memphis convention of the Alliance, held immediately following the election, a bitter fight erupted between Macune and Henry Loucks in the course of their contest for the national Alliance presidency. The decision turned not on personalities, but on a single and decisive political question: could an Allianceman in good standing oppose the People's Party? Macune soon discovered the answer was "No."

In an intricate series of maneuvers, the Texan—perhaps with memories of his dazzling speech at Waco in 1887—nominally withdrew from the race and then spelled out a new vision for a national cotton marketing program. He made no mention of his own sub-treasury plan, which, as orchestrated by William Lamb, had helped bring on the unwanted third party movement, and he ended by calling for farmer alignment with the Democratic Party in what he described as a "knockdown" blow at the forces of monopoly capitalism. Unfortunately for Macune, the Alliance "demands" and the Omaha Platform of the People's Party were identical; the delegates at Memphis knew well enough that the Democratic Party opposed the Alliance demands with consummate clarity. When a Macune partisan from Mississippi formally withdrew the Texan's name (was the plan to precipitate an emotional stampede to recall the founder to leadership?), a "red-headed delegate from Texas" jumped to his feet and quickly moved that nominations for president cease, leaving Henry Loucks of South Dakota as the sole candidate. The motion, promptly approved by the delegates, effectively ended Macune's career in the Alliance.

At the moment of the movement's most extreme vulnerability, the central figure of the agrarian revolt suddenly and permanently disappeared from the ranks, even as the crusade of the farmers moved into its climactic political phase. As if that were not bizarre enough, related to the sudden departure of the founder was another sequence of events that cast an ironic shadow over the abrupt culmination of the long tactical struggle between Charles Macune and the radicals of the Alliance.

William Lamb's personal political career—though not his organizational career—also had come to an end by the time of the Memphis meeting. The public standing of the red-headed organizer had, in fact, suffered a dramatic blow in the very week it reached its zenith with the formation of the Texas People's Party in August 1891. Absorbed in rallying the Alliance's growing left wing for the long-sought amalgamation with organized labor and the Colored Farmers Alliance in the new third party, Lamb had spent much of 1891 away from home, leaving the management of his farm to his sons and the operation of his newspaper, *The Texas Independent,* to his printing foreman. While Lamb was presiding over a meeting of the Texas Reform Press Association in early summer in East Texas, his foreman accepted a paid ad from an Alliance partisan that included a written attack on a local Montague County Democrat. Lamb's long years of Alliance activism had earned him the enmity of the hierarchy of the Hogg Democracy in Texas, and he arrived home to learn he had been sued for libel. As Lamb completed final preparations for the founding meeting of the People's Party in Dallas, the suit against him came to trial. The charge to the jury by the presiding Democratic judge defined the issues with such legal finality that Lamb angrily printed it in full in his own newspaper. The judge explained that neither actual nor punitive liability against Lamb could be diminished by the fact that he was not a party to the acceptance or publication of the ad. Neither were Lamb's public retraction and his offer to open his columns to the offended party to be considered by the jury as mitigating circumstances, again, either for actual or punitive damages. The jury did not award punitive damages, but it found Lamb guilty of libel and fined him $200.

The issue, of course, was not the size of the fine, but the political uses to which the conviction could be put. Lamb learned almost immediately how damaging the conviction was to his political life. At the founding meeting of the People's Party, comprised of Lamb's most intimate colleagues in radicalism, the delegates accepted his role as temporary chairman, but elected "Stump" Ashby as permanent chairman. The new party could not go before the electorate of Texas with an organizational posture that would permit its opponents to label it as a "party of slander." The libel case effectively removed Lamb, the red-headed apostle of farmer boycotts, from the leadership of the party he had been so instrumental in creating; it also prevented him from running for public office under its sponsorship. His personal standing in the Populist community was not affected, and he represented the fifth district in most of the party's internal functions throughout the Populist years. He was present in St. Louis in 1892 as Ignatius Donnelly read the platform preamble of Populism and watched quietly and perhaps with a touch of triumph as his old rival, C. W. Macune, was swept along into the third party by the forces Lamb had done so much to assemble.

But whether he was present in Memphis for Macune's demise later in the year is not certain; and whether he was the delegate who made the quick parliamentary motion that ensured Macune's exclusion from the contest for the Alliance presidency remains an interesting question. There is only the intriguing reference by a Memphis newspaper reporter to the motion made by "a red-headed delegate from Texas." It is not clear how many red-haired radicals were in the higher councils of the Texas Alliance. But it is clear that the long struggle of the two greenbackers—the economic radical Charles Macune and the political radical William Lamb—ended at the Memphis convention with victory for the third party men and the election of one of their own, Henry Loucks of South Dakota, as the order's new national president. It is also clear that both Macune and Lamb were casualties of their long political rivalry—one that culminated in the organizational consolidation of the National Alliance and the People's Party. Their duel, dating from the Cleburne Demands of 1886, had lasted six years. It personified one of the central ideological tensions at the heart of the agrarian revolt—

the power of an idea versus the power of inherited cultural loyalties.

7

The cooperative movement had built the Alliance to national stature, but in state after state in 1892, the cooperatives were strangling from lack of access to credit. The decline of the cooperatives presaged the collapse of the Alliance, for the dream of reform for farmers could now be given life only by the new People's Party. Only if the third party came to power and legislated the principles of the sub-treasury system into law could the farmers of the West and South escape the chattel mortgage companies and the furnishing merchants whose oppressive practices had generated the cooperative crusade in the first place.

The allegiance to "the party of the fathers" and to the party of white supremacy of such a central figure as Charles Macune underscored the deep-seated sectional and racial emotions brought into the open by the appearance of the new party of reform. The agony of Macune had its counterpart in every state Alliance and in every suballiance. This was true in the West as well as the South, for Populism challenged the "bloody shirt" as well as "the party of the fathers."

In the Populist strategy, the stultifying and self-defeating sectional memories of the American people were to be overcome by the power of an economic idea: the reform movement issued its challenge to the past in the modern language of the Omaha Platform. As Populists viewed matters, the grievances of "the plain people" had to be listened to by somebody, or the American idea of progress could simply come to describe the economic exploitation and cultural regimentation of millions. How to overcome the inherited sectional loyalties and political deference of an adequate number of Democratic and Republican voters remained the problem. At the end of 1892, the task of the reform movement was clear: it had to fully mobilize its own institutions in a concerted effort to teach the politics of Populism to the American people. Toward this end, the most important institution in the land, to many reformers, was not the People's Party. It was, rather, the National Reform Press Association.

8

Had Populism been nothing else, the Reform Press Association could stand as a monument to the democratic intensity of the agrarian crusade. The thunder of its great journals—*The Advocate, The American Nonconformist, The Appeal to Reason, The Southern Mercury, The People's Party Paper,* and *The Progressive Farmer*—was echoed in literally thousands of tiny, struggling weeklies across the nation. The famous journals struggled too; the common point of unity among the journalists of Populism was their shared poverty. Bereft of advertising support, the reform editors ran ads for books, frequently written by each other, and tried to survive on revenue from circulation. Unfortunately, many farmers literally could not afford to pay $1.00 or $1.50 a year for the good word of Populism. Usually they got the paper whether they paid or not, for the very reason for existence of the reform press was to "educate the people."

The number of Populist journals that materialized in the 1890's was well in excess of 1000. Kansas, Texas, Alabama, and Georgia led in journalistic insurgency, Kansas counting over 150 papers and Texas some 125. Alabama had approximately 100 reform papers and Georgia perhaps as many. Even the embattled third party cause in Louisiana counted about fifty journalistic supporters. Some Populist editors developed a talent for satirizing the local establishment. It sometimes proved dangerous, especially in the South. In Alabama, James M. Whitehead's use of ridicule in his Greenville *Living Truth* was "devastating"; his shop was broken into and the type scattered in 1892. The following year fire broke out, destroying the furniture in his rooms above the plant, though the press was saved. Whitehead continued his barbs without missing an issue. Another gifted satirist, Thomas Gaines of the Comanche *Pioneer Exponent* in Texas, was a recipient of attacks and threats of attacks, as was his family. The co-editor of "Cyclone" Davis's *Alliance Vindicator* was shot and killed in Texas. In North Carolina, Marion Butler's printing plant was burned.

The editors' remarkable personal experiences in coping with sudden violence engendered a sense of driving moral purpose that sometimes gave rise to lofty flights of description. Francis X. Matthews of the *American Nonconformist,* after returning from a crucial policy debate at the NRPA convention of 1895 at Kansas City, described his colleagues:

> This band of Populist editors, representing papers from all parts of the union, was . . . compelled to face such a storm of ridicule and vindictive hatred as seldom falls to the lot of men. They struggled against poverty in its most humiliating form, against ostracism and persecution and all uncharitableness. . . . The quality, tone, and contents of their papers steadily improved until today some of the most ably edited journals of the country are found in their ranks. . . . The note [they] strike is essentially national . . . Its appeal is to all people . . . regardless of creed, nationality, location, calling, or previous condition of servitude.

The sense of personal involvement so evident in the description by the *Nonconformist* editor was typical. Indeed, to most Populists—greenback theoreticians and dirt farmers as well as reform editors—the entire shape of American democracy seemed to have changed in the space of their own lifetimes. What was democracy when aggressive "captains of industry" could buy whole legislatures and keep the United States Congress in a perpetual state of genteel servitude? What was honest labor when ruthless structuring of the currency drove the price of farm products below the cost of production? What was thrift when high interest rates gobbled up farmland or when railroads made more money shipping corn than farmers did in growing it? Where was community virtue when bankers, commission houses, and grain elevator companies wantonly destroyed self-help farmer cooperatives? Where was dignity when farm women were forced to go barefoot and the furnishing man determined what a farmer's family could or could not eat? Where was freedom when the crop lien system was enforced by the convict lease system? What did the old virtues mean, in such a setting?

But however bad things might have looked to reform editors at the dawn of Populism, they soon got worse. In 1893

America plunged into a severe economic depression. The narrow organization of capital markets that derived from the doctrines of an inflexible gold currency precipitated the panic, which occurred during the annual financial squeeze caused by the autumn agricultural harvest; the panic in turn placed intense pressure on what all orthodox goldbugs regarded as the nation's "essential monetary reserves." To the utter dismay of green-backers, the sins of goldbugs seemed to multiply even as the reform movement gathered momentum. Grover Cleveland's Secretary of the Treasury was forced to sell massive gold bond issues through a syndicate organized by the J. P. Morgan Company, The effort unnecessarily saddled the country with a debt of over a quarter of a billion dollars, from which only bankers profited. Indeed, the government gained nothing from its frantic efforts, since bankers paid for the bonds with gold they had withdrawn from the Treasury! As one historian later put it, the Morgan syndicate "measured the emergency of the government with little mercy."

The depression inflicted great hardship. Millions knew genuine and prolonged privation. In the South and West there were increasing reports of starvation, both on the farms and in the cities. Industrial unemployment was so widespread that "tramps" became a familiar sight in every section of the nation. Marches on Washington by the unemployed gained increasing notoriety, especially since some of them originated in the mountain states and on the West Coast and involved "borrowed" freight trains as a primary source of transportation. The multitude became known as "Coxey's Army," after Ohio currency reformer and Populist spokesman General Jacob Coxey, who organized the first such demonstration. The peripatetic Henry Vincent of the *American Nonconformist* wrote a brooding account of this epic of industrial despair; he turned it into a book-length assertion of the need for the immediate rise to power of the People's Party. He called his book *The Story of the Commonweal.*

But, in the midst of the depression, the substance of the Populist message went not only from reform editors to their readers, but emanated from the poor themselves. The Populist farmers wrote both to their papers and to their spokesmen. Ignatius Donnelly received this letter:

I am ... one of those which have settled upon the socalled Indemnity Land ... now the great Northern. I settled on this land in good Faith Built House and Barn Broken up Part of the Land. Spent years of hard Labor in grubing fencing and Improving are they going to drive us out like tresspassers wife and children and give us away to the Corporations how can we support them. When we are robed of our means. they will shurely not stand this we must Decay and Die from Woe and Sorow We are Loyal Citicens and do Not Intend to Intrude on any R. R. Corporation we Beleived and still do Believe that the R. R. Co. has got No Legal title to this Land in question. We Love our wife and children just as Dearly as any of you But how can we protect them give them education as they should wen we are driven from sea to sea.

An Alliance leader explained to a Southern reform paper why his members could not pay dues: "Hundreds of good, hard-working men, true to the Alliance ... are staying at home, depriving themselves and their households of attending church, for want of decent clothing to appear in public. These people paid all they made to their merchant on their indebtedness, and are now, and have been, practically without a dollar. How can they pay dues?" In a letter edged with desperation, a farm woman living near Mendota, Kansas, wrote to Populist Governor Lewelling in 1894: "I take my pen in hand to let you know we are starving. ... My husband went away to find work and came home last night and told me that he would have to starve. He has been in 10 counties and did not get no work. ... I haven't had nothing to eat today and it is 3 o'clock." The sometimes angry, sometimes despairing dialogue between farm families and their Populist spokesmen provided the interior language of the agrarian revolt.

9

The nation's press did not push discussions of economic issues as defined by Populists. Ridicule was considered a more service-able tactic. It was sometimes augmented by a remarkable com-placency. Though, under E. L. Godkin, *The Nation* came to be hailed as "the best weekly not only in America but in the world," Godkin could do no better to account for the agrarian revolt than to describe it as "the vague dissatisfaction which is always felt by the incompetent and lazy and 'shiftless' when they contemplate those who have got on better in the world." More specifically, Godkin explained Western discontent to his readers: "a large body of farmers of that region are now really peasants fresh from Europe, with all the prejudices and all the liability to deception of their class."

Most metropolitan journals, however, did not bother to probe for explanations of popular discontent. They were more concerned with repelling its political expression, the People's Party. The patented all-season remedy for the third party was the politics of sectionalism. It had its uses, North and South. "A third party vote in Virginia," the Richmond *State* was certain, "is a vote for high tariffs and the Force Bill [for Negro voting rights]. Let our farmer friends make no mistake." While the Populist vice presidential candidate, Virginia's James Field, had to contend with the Richmond press in his native state, he encountered another brand of sectional politics on the Western plains. Commenting on Field's war service to the Confederacy, the influential Omaha *Bee* summarized the relevant issue: "Doubtless like most of his associates in that enterprise, he feels no regret at what he did, but being now a loyal citizen he wants to overturn the politics of the party which preserved the Union and substitute some of those which were promised in the event of the success of the Confederacy." Precisely which features of the Omaha Platform would have enthralled the leaders of the wartime Confederacy the paper neglected to specify. Meanwhile, in North Dakota, the state lecturer of the Farmers Alliance, campaigning as the gubernatorial candidate of the People's Party, was charged with being a "southern sympathizer." While rumors were circulated that he had been a "guerilla in Missouri" during the Civil War, the Republican candidate was widely billed through the state's press as a "loyal Union man." What all this had to do with government's response to the depression of the 1890's was far less clear than the emotional relationship of such memories to party regularity. Democrats and Republicans alike invoked the past to avoid the present.

10

In both regions of the country, Populists threw the Omaha Platform into battle in an effort to overcome sectionalism with the logic of greenback analysis. Dr. Stephen McLallin, the thoughtful editor of *The Advocate* in Topeka, was one of those who struggled throughout the 1890's to control his sense of outrage at the "sound dollar." "There is no more reason that the material in a dollar should have an intrinsic value equal to a dollar than that the yardstick should possess an intrinsic value equal to the value of the cloth that it measures. Money as such possesses neither length, nor thickness, and its only value consists of the fact that by law and custom it is the medium by which debts are paid and wealth exchanged." Though no modern economist would dispute that statement, it was, in the 1890's, a culturally inadmissible argument.

Populists like McLallin were fully aware that they were not only challenging old ideas which they found wholly obsolete in the new age of machines, but that in so doing they were engaged in a kind of cultural pioneering that made their proposals sound radical to the uninformed. The Omaha Platform required its defenders to be both judicious and culturally innovative—in order to overcome the sectional and cultural symbols available to the two major parties. In defending government ownership of the railroads or the sub-treasury plan Populists did not find it helpful to invoke Jeffersonian injunctions about "the government that governs least." Nor was Jacksonian preoccupation with hard money a notable aid in advancing greenback monetary analysis. Nostalgia for lost agrarian Edens, Jacksonian or otherwise, was therefore in short supply on third party hustings. Rather, the politics of the industrial age received the focus of attention. The rules of commerce had changed, and reformers knew it: indeed, the thought was at the very center of the Populist premise.

In the grip of this belief, Populists tested the intellectual flexibility of Gilded Age America. The substance of the third party's experiment in a new political language for an industrial society was the belief that government had fallen disastrously behind the sweeping changes of industrial society, leaving the mass of the people as helpless victims of outmoded rules. In an 1894 Labor Day speech Frank Doster, a Populist who later became the Chief Justice of the Kansas Supreme Court, decried the "fatal mental inability in both Democratic and Republican parties to comprehend the new and strange conditions of our modern industrial and social life, an utter inability to cope with the new and vexing problems which have arisen out of the vacillation of this latter day." The purpose of Populism, he said, was "to bring the power of the social mass to bear upon the rebellious individuals who thus menace the peace and safety of the state."

11

And so the strangely disjointed dialogue of the 1890's proceeded, the major parties relying on the nostalgia of sectionalism while the Populists offered the innovations of their democratic monetary and social program.

The forces of traditionalism were narrow in outlook, primitive in economic theory, and well protected by an enormous and passive constituency. Most importantly, the advocates of the "sound dollar" possessed all of the commanding heights in the culture—the nation's press, the universities, the banks, and the churches. Collectively, they had power.

The forces of reform, on the other hand, deployed several regiments of stump speakers, a thousand weekly newspaper editors, and a sizable constituency that carried strong but receding memories of the Alliance cooperative crusade. Collectively, they had hope.

It was not a balanced contest.

III
The Triumph of the Corporate State

My object is not to patronize the radicals by patting them on the head as "in advance of their time"—that tired cliché of the lazy historian. In some ways they are in advance of ours. But their insights, their poetic insights, are what seem to me to make them worth studying today.

Christopher Hill, *The World Turned Upside Down*

The truth of the matter is Taubeneck has been flim-flammed.

Henry Demarest Lloyd

7

The Shadow Movement
Acquires a Purpose

"A political party has no charms for me."

During 1892–95 the strongholds of the agrarian revolt gradually identified themselves through the cold statistics of elections from Virginia to Oregon. The "coming revolution," codified in the Omaha Platform after years of cooperative struggle and organizational development, had developed its strongest electoral base in three states—Georgia, Kansas, and Texas. But elsewhere across the nation during the same years the political style of the agrarian revolt underwent a variety of alterations that had very little to do with the doctrines so laboriously formulated at Cleburne, Dallas, St. Louis, Ocala, and Omaha.

Most easily identifiable of the new species was an effervescent "single-shot" Populism that had emerged in the Western mining states. Though sometimes producing victories for candidates who called themselves Populists, the Western mutation yielded a variety of "reformer" who proved difficult to distinguish from more familiar types calling themselves Democrats and Republicans. Indeed, in places like Nevada, where the cause of reform went by the name of the "Silver Party," a mention of the Omaha Platform was likely to extract little more from voters than blank stares. The doctrine of free silver was the overriding thought in the mining states of the West. Silver Republicans, silver Democrats, and silver Populists "fused" in a wondrous variety of ways to contest major party traditionalists on the only matter of interest to any of the participants—the increased coinage of the white metal.

Nevada was not unique. As Populists across the nation discovered in the course of surveying the election returns of 1894, the party of reform had begun to develop a dual personality. Not only in the mining states of the plateau region, but in a number of places through the old cotton belt and the newer Western granary third party organizational difficulties had led to what Alliancemen regarded as ideological "trimming" of the Omaha Platform. The political alternative that emerged as "Populism," particularly in some locales, bore little resemblance to the reform doctrines developed out of the cooperative crusade. As a matter of fact, with powerful help from outside the ranks of Populism, a shadow movement began taking precise political form.

2

The phenomenon was clearest in Nebraska. The movement that materialized there was less a new body of ideas than a loosely floating faction of the familiar low-tariff Democratic Party. Indeed, the so-called "Independent" movement that came into being in the midst of "fusion" politics in 1890 had subsequently ritualized the practice of Populist-Democratic cooperation into a predictable habit of thought. Nebraska independents could not even claim to be the initiating force in this amalgamating process; Democrats were. And the most zealous Democrat was William Jennings Bryan, the young Congressman elected in 1890. Though Bryan had rather traditionally stressed the Democratic issue of low tariffs in his 1890 campaign, the proliferation of agrarian demands on the "financial question" induced him, during his extremely close re-election campaign of 1892, to begin calling for "free and unlimited coinage of silver," a slogan that harkened back to the days of "The Crime of '73," some twenty years earlier. While Bryan had at best only a shaky grasp of the intricacies of the monetary system, he knew that silver coinage avoided awkward moral questions about a "fiat currency" and an "honest" dollar. Bryan conceded in 1892 that he did not "know anything about free silver," and he added, cheerfully, "the people of Nebraska are for free silver and I am for free silver. I will look up the arguments later." For a major party politician in search of an issue through which to survive in an era of reform agitation and hard times, silver coinage seemed ideal. As one writer put it, Bryan "looked up the arguments" and made a fiery free silver speech in the Congress in 1893. He immediately acquired new friends among the silver mineowners and they promptly printed almost a million copies of the speech for distribution. The action signaled the arrival of a new element in Gilded Age politics—the American Bimetallic League.

Because of its enormous impact in giving genuine political meaning to the Nebraska shadow movement, the story of the silver drive constitutes the essential background to the unfolding of the changed national politics of 1895–96. While it was eventually to affect the cause of reform in a decisive way, the true center of gravity within the shadow movement rested outside the People's Party.

Though the American Bimetallic League had been formed by silver mineowners as early as 1889 and held its first non-partisan convention in St. Louis that year, it was not until the depression of 1893 had placed severe pressure on the nation's gold reserves that silver coinage began to have serious possibilities as a national issue. As the depression deepened, President Cleveland's orthodox monetary beliefs induced him to take extraordinary measures to preserve the gold standard, including the expensive, unpopular, and widely publicized gold bond issues. Clearly, the best way for an embattled Democratic politician to dissociate himself from the goldbug policies of Cleveland lay in becoming a public advocate of the white metal. In 1893 the silverites hired a publicist named William H. Harvey to supervise their lobbying activities in the Midwest. Harvey himself began writing inexpensive silver tracts aimed at a mass audience and marketed through advertisements in the Democratic and third party press. The first one, "Coin's Handbook," enjoyed a modest sale.

The silver interests also broadened their geographical base. The Omaha *World-Herald* was purchased in 1894, and William

Jennings Bryan was installed as its editor. The silver men put
in their own managment, leaving Bryan free to campaign for
silver while contributing an editorial or two each week. Other
connections were also forged. Such newspapers as Edward
Carmack's Memphis *Commercial-Appeal* began to demonstrate an
astonishing preoccupation with the silver issue. So did the
Washington *Evening Star,* thus providing a base of support in the
nation's capital to augment the outposts in the South and West.
The American Bimetallic League was reinforced by the "National
Bimetallic Union" and the "Pan-American Bimetallic Association," and by the emergence of scores of local and regional silver
"clubs," silver "unions," and silver "leagues." In the summer of
1894, the Omaha *World-Herald* announced Bryan's candidacy
for the Senate at the "request" of the executive committee of a
new group called the "Nebraska Democratic Free Coinage
League."

Silver influence popped up in a number of places. Scarcely had
the Omaha Platform been unveiled in 1892, in fact, than Herman
Taubeneck, the national Populist chairman, began writing a
select group of third party politicians of his interest in procuring
some campaign contributions from Western mining interests.
Meanwhile, Charles Macune's long-time editorial assistant, Nelson Dunning, inaugurated *The National Watchman* an an organ
for the new third party. Coordinate with silver lobbying inside
the Democratic Party in 1894–96, the *National Watchman* began
to prosper. For a Macunite, Dunning seems to have found it
surprisingly easy to dispense with the greenback doctrines woven
through the Omaha Platform. Dunning, in fact, became as
dedicated to the new cause of silver coinage as Carmack's
Commercial-Appeal or such robust new journals as the Chicago-
based *National Bimetallist* and the Washington *Silver Knight.* The
latter journal was edited, nominally at least, by Nevada's silver
Senator, William Stewart, a Republican.

Though such diverse support was important to William Jennings Bryan in his new role as one of the nation's spokesmen for
silver, his political success at home in Nebraska was directly
traceable to the ease with which he was able to work with the
representatives of the so-called "Independent" or Populist move-

ment in the state. Indeed, Bryan had a close friend at the very apex of the new party movement.

By a curious and revealing twist of circumstance, the titular leader and acknowledged spokesman of Populism, Nebraska-style, came to be neither Charles Van Wyck, the disenchanted Republican antimonopolist, nor John Powers, the well-meaning agrarian spokesman, but a friendly, apolitical, and previously unknown small-town lawyer named William V. Allen. As a United States Senator from Nebraska, Allen became the archetype of a strange new breed of third party politician who was a Populist in name only.

William Allen actually had been a typical Republican until 1891. Following his late conversion he was never able to piece together in any coherent fashion either the rationale behind the Omaha Platform or an explanation of how such tenets came to find their way onto the Nebraska political scene in the first place. Not only the doctrines, but the very origins of Populism seemed to baffle him. Confusing the 1888 and 1889 Alliance platforms with the document drafted at Ocala in 1890, he cheerfully described the Ocala gathering to a New York editor as "the convention of political and agricultural reform elements" that "met a year or two before" his own conversion "in 1890." Allen's innocence of the agrarian movement for which he spoke was not without its immediate political meaning: whatever others might say of Populism, the Nebraskan made it clear he was not a "radical party man." He consistently avoided not only green-back monetary analysis, but also such other doctrines organic to the Populist platform as government ownership of the railroads. As Allen's public statements made clear, the entire subject of the Populist platform tended to fill him with caution. He would not "start off with fiat money," he explained, because paper money should have "its interchangeability maintained" and "kept as good as gold and silver." As firmly as any gold monometallist, Nebraska's Populist Senator thus endorsed a currency of ultimate redemption. Equally beyond his purview were such Alliance concepts as the democratic vision of basic realignment of the American party system. Far from seeing the People's Party as a new political institution to bring representation to the "industrial millions," Allen asserted that "a party should be held no more sacred than a man's shoes or garments, and that whenever it fails to serve the purposes of good government, a man should abandon it as cheerfully as he dispenses with his worn-out clothes." Confessing that he had taken "very little part in politics" in the past, Allen summarized his credo with the thought that "a political party has no charms for me outside of what it can accomplish conductive to good government." A bit more specifically, he said that he was "not a radical" on the tariff and "in no sense a socialist" on the labor question. Amid the flurry of disclaimers, it was not always easy to isolate precisely what the Nebraska lawyer did endorse. But no one questioned his sincerity. In essence, William V. Allen stood for honest, conservative government. He said as much in a summary statement for the *Review of Reviews:*

> If my views could be enforced in this country I would purify state and national legislation. I would not suffer a man to become a member of either branch of Congress or of any state lawmaking body who had pecuniary interests which might be materially affected by legislation. If he wanted to become a member of Congress he would have to put aside his own precuniary interests for the time, so that he might be said to stand as an impartial judge in the determination of any case that came before him. This is foreshadowed, you see, by the bill I introduced the other day entitled "A Bill to Preserve the Purity of National Legislation, and for other Purposes." The bill is imperfectly drawn, but I shall hereafter redraft it with more care. I do not suppose it will get through this Congress, and perhaps it will never get through.

Manifestly, William Allen was cut from different cloth than Populists, as the latter were perceived either by most of the nation or by most Populists. Yet Allen authentically represented the style of Nebraska Populism, both in his beliefs and in the very manner in which he came to be the state's third party Senator.

Allen gained his Washington office in 1893 as a result of a confused bit of fusion in the Nebraska legislature, which had an interesting assortment of Republicans, Democrats, and "Independents." The Republicans had slightly less than a legislative majority over their combined opposition, and balloting for the United States Senate seat was lengthy, confused, and threaded with rumors about back-room deals. After twelve ballots the Republican senatorial candidate, who was, appropriately enough, the general solicitor for the Union Pacific Railroad, almost gained a majority, but the anticipated support of five Gold Democrats failed to materialize at the last moment. Holding the balance of power, the Gold Democrats sought bigger game. For their part, the Nebraska Independents attempted to win the same conservative Democratic support, and to that end they organized coalitions behind three successive candidates—each a bit more acceptable to conservatives than his predecessor. The third political specimen offered to the Democrats was William V. Allen.

After mobilizing all possible Democratic and Populist support for their man, the fusion strategists discovered that they were still one vote short of a majority. The five Gold Democrats then confronted the Republicans with a narrow choice: either the G.O.P. would support a conservative Democrat—such as that great and good friend of the railroads and the sound dollar, J. Sterling Morton—or the Democrats would vote for Allen. The conservative *Nebraska State Journal* reported that the Republicans were buckling to this pressure, but at the last moment they decided to give one of their own a final chance at the senatorial laurel wreath. With that, the Democratic conservatives made good their threat, voted for Allen, and, in this manner, gave Nebraska a spokesman for reform in the United States Senate. The Republican Omaha *Bee* expressed its surprise and pleasure at the final result, particularly in light of the rather low form of knavery that had characterized the balloting. The paper, which had authored a sprightly series of bloody-shirt attacks on Populists throughout the campaigns of 1890 and 1892, judged Allen to be "well balanced, broadminded and conservative." He stood, said the paper, "head and shoulders above any other man proposed by the Populists in point of ability and honesty of purpose."

The easy harmony between such hard-money, low-tariff Democrats as William Jennings Bryan, such committed Republican papers as the Omaha *Bee*, and such moderates as William Allen adequately defined the nature of Nebraska Populism. Virtually issueless, the third party in Nebraska was simply, decently, and solely dedicated to "honest government."

The moderate coalition worked out elaborate fusion arrangements in congressional districts and even at the local level. The supremely confident Republicans increased the stakes by naming a gubernatorial candidate who was not only clearly identified with the railroads, but was also tainted by some unsavory financial dealings. Times remained hard, a State Relief Commission Report referring to the "utter destitution of the industrious thousands." During 1890–95 well over half a million high-interest chattel mortgages were recorded on Nebraska farmers, who, because of the contracted currency, had no other means of obtaining money.

Despite the pressure from the financial system, as well as advantages accruing from the variety of fusion arrangements and the rare benefit of a divided metropolitan press, the moderate coalition again met clear-cut defeat at the hands of the Republicans in 1894. Only one Republican statewide candidate, the scandal-marred gubernatorial aspirant, failed of election. Republicans won 75 per cent of the seats in the state senate and over 70 per cent of the house.

"Railroad candidates" were safe in Nebraska; the mild lesson for the Republican hierarchy produced by the campaigns of 1893 and 1894 was that only "scandal-tainted railroad candidates" were vulnerable. Of the total vote in the state, the People's Party received but 34 per cent in 1894. By every criteria that bore on the task of "educating" voters on reform issues, the third party leadership in Nebraska had proved remarkably incapable of reaching its natural constituency. The Nebraska People's Party languished in complete ideological homage to the Democratic Party; the Omaha Platform was clearly an irrelevant piece of paper. Indeed, at campaign time its existence only created awkward problems for the cause of harmony and fusion. Silver had therefore found not one spokesman, but two—Populist Allen and Democrat Bryan. They shared identical political perspectives. The fate of greenback doctrines in Nebraska was wholly unique in the tier of states from North Dakota to Texas along the farming frontier of the Great Plains.

3

The emergence of such an anemic version of Populism as the Nebraska shadow movement clearly revealed the depth of the cultural barriers to reform faced by the new movement. Indeed, the tone and the substance of the third party varied considerably across the nation. A review of Populism in Alabama reveals the tensions generated when the cause of reform swerved from its announced intention; and a review of Texas and Kansas Populism uncovers still other hazards awaiting the reform movement when it did not swerve.

In Texas what the third party got for its dogged efforts in 1894 was largely the empty fruits of stolen victories. The principal post-election activity of Populist strategists focused on discovering a way to win contested election lawsuits in a state legal system dominated by Democratic judges. "Harrison County methods," like the "Richmond County methods" of Georgia, had become the primary campaign technique of Democrats. The legal system, too, it turned out, was hierarchical. In any event, the 160,000 votes "officially" counted for the third party verified the growing ability of Populists to recruit from the original Alliance constituency. The ideological strength of Texas Populism lay in this fact—that the most available potential recruits had been old Alliancemen who were not necessarily locked into one or both major parties in ways that called for "fusion" arrangements. The tactical course, therefore, was self-evident; as the Southern *Mercury* never tired of putting it, Populists merely needed to "continue the patient program of education" in order, ultimately, to bring all 250,000 of the Alliance brethren "home" to Populism. The fact that, in 1894, some 90,000 still remained loyal to "the party of the fathers" testified to the enormous power of sectional and racial memories, particularly in the states of the Old Confederacy. By 1894, in any case, one thing was clear: the Omaha Platform was the uncontested symbol of the Texas movement and the party there was busily engaged in "fighting it straight." There was at least one strong reason for optimism—the wagon trains to Populist encampments were getting longer.

In Kansas the agrarian crusade had materialized in 1890 as an extraordinarily passionate happening, and the political hopes of the huge Kansas Alliance had been given broadly radical definition by the hundreds of greenbackers who had helped carry the movement to the four corners of the state. The heady outlook of 1890, diminished by the defeats of the third party in municipal elections in 1891, indicated that the transition from Alliance politics to Populist politics had apparently jolted a number of temporary recruits back to their ancestral homes in the major parties. The chastened third party in Kansas accepted fusion with Democrats in 1892, though the procedure was for the most part carefully formulated on the basis of the Omaha Platform. The third party's constituency in Kansas insisted on this precaution, which was overwhelmingly confirmed in the Populist state convention in 1892. The fusion candidates won that November, but it became obvious that fusion carried dangers beyond the matter of ideology. Democratic participants in the coalition expected patronage as their reward for sacrificing their principles, and the "practical" wing of the third party—Populist officeholders—moved to accommodate them. Both Populist Governor Lewelling and third party legislators were aware that the margin of votes brought to the coalition by Democrats, while small, had been decisive. Without the Democratic votes the Populist ticket would have been defeated. The patronage policy decided upon by the Populist political leadership culminated in the election by the Kansas legislature of a new United States Senator—a Democrat. The move deeply troubled a substantial number of Alliance leaders and others among the third party faithful. An organized uprising from below soon materialized that made fusion all but impossible to consummate in 1894, however eagerly the party's politicians might have desired it.

Analysis of the succeeding campaign was rendered a bit obscure by the issue of woman suffrage—which the Populists supported and the Republicans opposed—but the result was abundantly clear: Simpson and the other Populist luminaries all went down to defeat in 1894. The Republican Party elected its entire state ticket, seven of eight Congressmen, and 91 of 124 state legislators. The third party's share of the total vote provided the most sobering statistic of all, however, for it revealed institutional stagnation. The maximum allegiance of Kansas to the Populist cause stood at 39 per cent of the electorate—roughly the same number of people who had been reached by the Alliance cooperative movement. Meanwhile, the Democratic Party, with less than 30,000 adherents, had seemingly been all but destroyed in Kansas. The Republican Party, on the other hand, had emerged as the dominant political institution of the state; it possessed an absolute majority over its combined Populist and Democratic opposition. As 1894 approached, it was apparent the third party had reached some sort of crossroads in Kansas.

Third party fortunes in Alabama illustrated other problems confronting the insurgent democratic movement. In Alabama, reform groped through the thickets of sectionalism. The state's foremost reform spokesman, Reuben Kolb, was a hesitant politician. Despite the grotesque vote frauds that characterized conservative Alabama politics in the 1890's, never at any time was he able to make the ultimate psychological break from the party of the Confederacy. The democratic clarity of Alliance demands had, in fact, kept Kolb a bit off balance ever since the St. Louis meeting of 1889. Macune's sub-treasury plan, greenback doctrines generally, and such issues as government ownership of the railroads appeared almost (though not quite) as unnerving to Kolb as they did to William Allen and William Jennings Bryan in Nebraska. The difference was the enormous pressure organized Alliancemen could place on an Alabama "leader," a force not present in Nebraska. A spring day in 1892 when Clay County farmers greeted with dead silence Kolb's declaration of loyalty to the Democratic Party was symptomatic of the intensity of the problem confronting him in that year of ultimate political decision. Though such democratic intensity from below eventually pushed Kolb into an insurgent campaign, the vote frauds of 1892 provided him with a new way to avoid the radical clarity of the Omaha Platform. His campaigns now centered around a call for "a free ballot and a fair count." Granted that this objective (like the goal of "purification" in Nebraska) had a certain relevance if the idea of democracy was to gain credibility in Alabama, the extreme emphasis that Kolb personally placed on the issue in the federal elections of November 1892 diverted attention from basic third party goals. Nevertheless, the divisions that materialized in the electorate were real enough. The sheer magnitude of the bitterness, abuse, and violence that attended the third party's struggle in Alabama verified the growing challenge to conservative rule. The alarming fact that explained the conservative reign of terror in Alabama was a simple one: "the party of the fathers" had lost the struggle to hold the loyalty of the state's white farmers. The Alliance movement had been built on twelve-cent cotton that left farmers no hope of escape from the crop lien system. In 1894 the price sank to five cents. A great deal was being asked in the name of party loyalty: did the farmers seriously have to accept the possibility of starving in order to protect white supremacy?

Alabama's all-Democratic congressional delegation came to the conclusion that too many of their constituents were not measuring up to this new test. Indeed, a discernible panic came over orthodox Democrats. The first to break was William Denson, Congressman from the impoverished seventh district. In 1894, he announced a personal decision to go off the Cleveland standard. He informed his colleagues that he would seek re-election as a Populist. It soon developed that the same option was being seriously considered by a number of them! A mass defection was, in fact, seriously discussed. Such a dramatic course not only promised to ease the pain of transition; it might have a certain political selling power back home. That Populism had been denounced by each of them as a threat to every sacred tenet in the Southern creed admittedly posed a certain problem; on the other hand, there was the increasing certainty of defeat to consider. The Alabama Congressmen agreed to put their proposal to John T. Morgan, the state's influential Senator. A stout Democratic regular who could launch eulogies to the Southern heritage with the best of the silver-haired colonels, Morgan repeated all the familiar litanies of railroad enterprise, white supremacy, and honest money—in approximately that order. His advice to his congressional colleagues, however, indicated the nadir to which the Southern Democracy had sunk. Turn Populist? "If I were younger, I would do it myself," he said equably. But, he sighed, he was too old to become an insurgent. He was now a free silver man himself, he added, and he would tell the Alabama electorate so in no uncertain terms. The people needed relief—he was certain of it! He showed his anxious guests to the door and left them to ponder their own course in the quiet of their own consciences.

The organized defection of the Alabama delegation collapsed with this pledge of non-support, though the subsequent elections confirmed how badly matters had deteriorated. Of the nine Democrats in the delegation, only three legally survived the cataclysm of 1894. Two more were "counted in," but the remaining four could not be saved—not even by fraud. The party of white supremacy again demonstrated its remarkable electoral appeal to black voters, amassing margins of forty to one and even fifty to one in the black belt.

The farmers of Alabama had been duly educated on the issue of a free ballot and a fair count; only their conservative opponents failed to grasp the merit of this plank in the platform of Reuben Kolb and the Populists. Clear to all, however, was the certainty that the gold standard no longer provided adequate support for serious Democratic politicians. As Democratic Senator Morgan had forecast, "free silver" offered the last handhold on the cliff. If the Democrats did not grasp it, the entire state of Alabama would plunge into the murky waters of Populism.

The salability of the free silver tactic obviously depended on the electorate's understanding of the money issue. In Alabama, Reuben Kolb's stewardship had left some promising openings for Democratic regulars. The proposition, in any case, would be tested in 1896. For the time being, and depending upon which third party orator one used as a guide, Alabama Populism emphasized radical greenbackism, honest elections, or both.

4

The elections of 1894 proved difficult for Populists to analyze. People's Party strategists discovered in that year that while the reform movement had made sizable gains in the South, it had taken unexpected losses in the West.

In Georgia, Tom Watson, though again robbed of election, led a still growing third party to win 45 per cent of the total state vote—and this in the face of returns even Democrats conceded to be fraudulent. Populists counted over a dozen contested elections from Texas, Alabama, Georgia, and Louisiana as testament to both the strength of reform and corruption of the opposition. Yet the resurgence of the Republican Party in the North and West was shocking, even though it testified to the intense voter dissatisfaction with Grover Cleveland's administration in time of depression. While many voters, particularly in the South, had turned to the third party, even more voters in the West seemed to have turned to the Republican Party. The electorate behaved in strangely volatile ways, Populists told themselves, avoiding speculation that the sizable shifts of 1894 might indicate a different kind of party realignment than the one reformers anticipated.

Whatever the long-term implications, Populists were of at least two minds about what to do. In Kansas, Jerry Simpson nursed the bruises from his unexpected defeat and decided that Populists needed to cooperate more with Democrats. In Colorado, where both the third party and the Democrats had been roundly defeated, other Populists began to consider the same possibility. In the Republican-dominated West, fusion would necessarily be with Democrats, of course, and it could be achieved on the issue of free silver. The idea of a Populist-Democratic coalition, institutionalized since 1890 in Nebraska, thus acquired growing support in 1895 in a number of places in the West. However, the suggested tactic appalled greenback reformers, especially in the South, where elections had become a tense, bitter, and even violent battleground between Democrats and Populists. Tom Watson thought the silver issue to be shallow, and dangerous to the reform party as well. Tom Nugent, in the Southwest, reached the same conclusion, as did other prominent Texas Populists. The reform press, both in the West and South, overwhelmingly shared their opinion, as did Ignatius Donnelly in Minnesota, Davis Waite in Colorado, and George Washburn, one of the party's few prominent spokesmen in the East.

A month after the 1894 general election had brought proof of the party's losses in the West, national party chairman, Herman Taubeneck of Illinois, declared himself: the People's Party had to jettison the Omaha Platform and "unite the reform forces of the nation" behind a platform of free silver. The shadow movement had not only acquired a precise form, a vague but determined commitment to "free silver," it has also acquired status in the personage of the Populist national chairman.

To veteran Alliance greenbackers, the news was appalling. The *Southern Mercury* reacted promptly and violently: the Omaha Platform was the basis of Populism and could not be jettisoned without abandoning all that the reform movement meant. Trade the "coming revolution" for a silverized currency of ultimate redemption? The idea was "disappointing and disgusting," said the paper. Such a course would not only betray the Omaha Platform, it would scuttle every Alliance declaration all the way back to the Cleburne Demands.

But the party's chairman and a number of its prominent politicians—Allen of Nebraska, Simpson and Lewelling of Kansas, and the new coalition Senator from North Carolina, Marion Butler—all seemed to be quite serious. In response, the party's rank and file, centered in the old Alliance movement and buttressed by an overwhelming majority of the editors of the reform press, rallied to defend their cause. Suddenly, and with considerable passion, the People's Party moved headlong into its final crusade—one in which Populism encountered its shadow movement in a tense struggle to define the meaning of American reform.

8

The Last Agrarian Crusade
The Movement vs. The Silver Lobby

"We propose to stand by the demands of the industrial people!"

It would be much too facile to portray the fierce internal struggle within the People's Party in 1895–96 as a simple clash between conservative "fusionists" on the one hand and radical "mid-roaders"* on the other. For one thing, it was really a struggle between democratic politics and conventional hierarchical politics. For another, the decisive ingredient in the struggle did not emerge from within the ranks of Populism at all, but rather from economic and political groups completely outside the People's Party.

Fusionists had a common characteristic. Almost all of them held office, had once held office, or sought to hold office. They saw the future of the third party in immediate terms. Some, like Senators William Allen and Marion Butler, exhibited no visible qualms over dispensing with virtually the entire Omaha Platform; others, like James Weaver and Jerry Simpson, had green-

---

* The term "mid-road" derived from the 1892 campaign, in which Populists saw their new party as charting a new multi-sectional course between the Republican Party of the North and the Democratic Party of the South. The path of third party radicalism was thus "in the middle of the road" between the sectional agitators in the two old parties. In both the South and West the phrase was widely used by Populists in 1892 to describe their intention to "fight it straight" by opposing both major parties. The term could not, of course, be used to describe the politics of the shadow movement in Nebraska—not even in 1892.

back pasts. All wanted to win—and at the next election. For their part, mid-roaders had a more recognizably shared past, derived from an early identification with the Alliance movement. They also tended to be from states that possessed strong third parties or growing third parties. Tom Watson of Georgia, Thomas Nugent of Texas, and E. M. Wardell of California were rather typical of this group.

At bottom, the third party's internal struggle was a contest between a cooperating group of political office-seekers on the one hand and the Populist movement on the other. The politicians had short-run objectives—winning the next election. In contrast, the agrarian movement, both as shaped by the Alliance organizers who had recruited the party's mass base of partisans and as shaped by the recruits themselves, had long-term goals, fashioned during the years of cooperative struggle and expressed politically in the planks of the Omaha Platform. While the movement itself had a mass following, the only popular support that the office-seekers could muster within the third party itself was centered in those regions of the country which the cooperative crusade had never been able to penetrate successfully. In some of these regions, notably the industrial East and the upper Midwest, the third party had only a tiny following. In others, especially in the mining states of the West, Populism had drawn its adherents not through the experience of the cooperative crusade, but because, simply enough, one of the elements of monetary reform advocated by the third party concerned the free coinage of silver. Miners wanted free silver not because it meant a "new day for the plain people," but because it meant jobs for themselves. Finally, in one state— Nebraska—the third party following was neither tiny nor organically silverite. Rather, as a shadow movement essentially created outside the framework of the cooperative crusade, the Nebraska Party was not organically anything. Therefore, that state, like the non-Populistic states of the East and upper Midwest and the mining states of the West, was ripe for the kind of third party strategy harmonious with the short-run needs of the party's office-seekers. In sum, where the agrarian movement was strong and growing, the politics of the movement was intact;

but where the movement had never sunk genuine roots, or had become stagnant, the third party's political stance was co-optable.* In general, therefore, the contest between Populism and its shadow form in 1896 arrayed the politics of a people's movement against conventional electoral politics. More specifically, it arrayed the democratic politics of the movement culture against the hierarchical politics of the received culture.

But Populists were not the only ones driven to serious reflection by the turbulent politics of the Gilded Age. Democrats were also pressed toward wholesale reevaluations by the fury of the agrarian revolt. The old party had taken massive losses in 1892 in the West, and the defections in the South reached such tidal proportions in 1894 as to imperil the very future of the party in its "Solid South" heartland. In both regions, Democratic regulars found Grover Cleveland's dedication to "sound money" to be an expensive obsession, one that threatened the destruction of the party as a national institution. After the ominous Southern vote for the People' Party in 1894, the prospect of seeking re-election in the presidential year of 1896 while yoked to a ticket headed by a "Gold Democrat" terrified Democrats from the Ohio River to the Gulf of Mexico. A very real prospect existed that the entire South might fall to Populism and wreck long-standing Democratic political careers by the hundreds. Similarly, everywhere west of the Mississippi River the Democratic Party had virtually ceased to exist as a credible institution. In Kansas the Democrats polled less than 30,000 votes out of 300,000 in the general elections of 1894. In Nebraska the party mustered less than 20,000 supporters in a voting population of 200,000. In Minnesota, the Dakotas, Colorado, Montana, and Oregon the story was the same—the doctrines of the gold standard were eating away the very roots of the Democratic Party. To anguished Democratic regulars the evidence was overwhelming: unless something were done to alter traditional Democratic politics, the basic party realignment radicals dreamed of would be confirmed in 1896—at the expense of the party of Jefferson and Jackson!

---

* For a discussion of the underlying structural causes of this vulnerability see pp. 139–44, 215–22, and 298–310.

2

Among the Populists, this state of affairs generated two dia-
metrically opposed analyses. The first appealed to third party
politicians with immediate short-run objectives. It went substantially as follows: The absence of a potential Populist plurality in
the West, confirmed by the elections of 1894, made fusion with
Democrats a necessity if the reform cause were to have a firm
basis upon which to challenge Republican hegemony in 1896.
The discomfiture of Democrats was a blessing, for it created
"Silver Democrats" by the wagonload. All the People's Party had
to do, therefore, was to concentrate on the silver issue, unite all
dissident forces, and thus bring the "coming revolution" to a
solidly contending position in 1896. Of course "a coalition of
the reform forces" could not be made on the basis of the Omaha
Platform, as the prospective "catch" of disillusioned Democrats
and the lesser number of "Silver Republicans" would not take
the bait of so radical a program. But a one-plank Populist
platform of unlimited silver coinage would create a winning
national coalition. People had memories of silver from the days
preceding demonetization. They were used to silver. The white
metal was not "radical." It was a vote-getter. It would bring the
reform movement to power. Such was the case for Populist-
Democratic fusion.

A small but prominent group of Populist politicians desperately seized the silver issue and argued for its immediate adoption
as third party policy. As they saw the practical options, the
People's Party had done its best to convince American voters of
the need for the Omaha Platform. After four years the party
had gained a following of anywhere from 25 to 45 per cent of
the electorate in twenty-odd states. These facts could be summarized in one word—defeat. Politically, the mathematics of the
situation were fatal. The People's Party needed to broaden its
base or see the cause of reform die completely. One did not
have to be a Populist politician to accept the fact that reformers
had to be in office in order to enact reforms. The whole matter
thus constituted an open and shut case of recognizing political
reality. To the group of political aspirants who asserted these
arguments, the fusionist cause also added large numbers of

rank-and-file party members in the Western mining states. Miners in the West did not know or care about chattel mortgages, crop lien systems, land loans, commodity credit, or the relevance of explanations about the commodity value of debts in a contracting currency. Free silver meant full employment in the Western mining centers. Unlimited coinage meant the opening of marginal mines. It meant Western business expansion. It meant prosperity.

But, as veteran Populists viewed matters, two things were wrong with the free silver assessment, and they both pointed to one conclusive reality: it would destroy the people's movement. First, in terms of its long-range political effect, one-plank free silverism threw away one of the third party's basic goals at the very moment it was coming within reach. Party realignment was an important objective in itself, because it was a precursor of larger objectives. Since both old parties were in harmony with monopoly, it was necessary to restructure the party system in order to restructure the nation's financial and economic system. That was why the People's Party had to navigate between the sectional barriers of the two old parties and recruit voters from both by hewing to "the middle of the road" and "fighting it straight." Indeed, that was what the Omaha Platform was all about. Even granting the highly debatable proposition that a free silver platform might achieve permanent party realignment, no thoughtful person had any illusion that it would address the underlying economic objective. In fact, by so sharply defining the limits of reform, a victory on the basis of a silver coalition would actually preclude the possibility of attaining significant alterations in the prevailing forms of the American monetary system. Free silver did not alter the existing banking system, nor did it end the destructive privilege national bankers enjoyed through their power to issue their own bank notes on which they gathered interest. Mid-roaders regarded this post-Civil War folkway as a veritable license to steal. Silver coinage elevated the folkway into a system; worse, by deflecting debate from the banking system itself as the proper object of reform, free silver actually undermined the greenback cause. What did silver coinage have to do with a flexible currency, for example? While "unlimited coinage at the ratio of sixteen to one" might end the contraction of the currency—if the silver mines held out long enough—it did not provide a flexible monetary system keyed to population expansion and industrial growth. Indeed, it did not even alter the metallic basis of the currency—for silver at a sixteen to one or at any other ratio was a coin of ultimate redemption. Though silverites seemed unable to ponder the monetary system long enough to think the matter through, hard-money inflation that left the banking system undisturbed utterly failed to address the real needs of the nation's producing classes. Anyone could see that free silver provided neither land loans nor commodity loans. Veteran Alliancemen and reform editors summarized their case by reminding party politicians that free silver was a pallative that would ward off the liberating triumph of a fiat currency. In terms of a decisive monetary breakthrough under capitalism, free silver had therefore become a step backward.

The second objection to Populist-Democratic fusion turned on the need to rescue the democratic process from the permanent corruption that was rooted not only in the monetary system, but also in the power of large-scale capital to shape the substance of American politics. In not the slightest way did silver address the accelerating movement toward industrial combination. As John D. Rockefeller had conclusively demonstrated in the course of creating the Standard Oil Trust, railroad networks were a central ingredient both in the combination movement itself and in the political corruption that grew out of monopoly. Silver coinage utterly sidestepped the whole matter of government ownership of the railroads. What were silverites proposing—that railroad lobbyists begin buying state legislatures with silver coin instead of gold? Would that noticeably improve the level of ethics in the United States Senate? Mid-roaders could summarize all their arguments by asserting that the ultimate monopoly was the "money trust," a banking system of private plunder anchored in a metallic currency and assured of power because it owned both "sound-money" parties. "Free silver" was irrelevant to all of these realities; it did not address corporate concentration.

On intellectual as well as political grounds, then, both the cooperative and the greenback heritages of the People's Party stood in the way of the ambitions of one-plank free silverites.

Allied with Alliancemen was an overwhelming majority of the
editors of the reform press. Their commitment to the reform
movement had been an intellectual one, engendered by the
modern sweep and practicality of the Omaha Platform. Because
of the personal hardships they had endured, including economic
hardships, reform editors scarcely looked upon the ambitions
of fusionist politicians with tolerance. To the editors, fusionists
seemed to be little more than a clique of self-interested oppor-
tunists who would sell out the cause of the people for another
term in office.

These Populist mid-roaders offered a "practical" case, too.
They believed the People's Party merely needed to keep its eye
on basic reform purposes. In long-run terms, the third party
cause had come along in fine shape. Greenback doctrines were
no longer confined to a narrow tier of frontier counties in the
West; now they permeated a great mass movement which in
turn had created a national third party. The party had deep
and still spreading roots, all the way from the tenth district of
Georgia on the Atlantic coast to the working class precincts of
San Francisco. The crucial political reality lay in the fact that the
national Democratic Party had been fatally undermined by the
politics of Populism. As constituted, the Democratic Party had
all but disappeared from the plains states westward to the Pacific
Coast, and it was tottering toward collapse across the South. To
mid-roaders, the new political reality was plain; the mad rush
of Democratic Congressmen to embrace free silver merely
verified the power of Populism. Under the circumstances, all
reformers had to do was to steer a steady course and hold to it.
The clinching argument to mid-roaders was the self-evident fact
that Democrats had no saving expedient. Conceivably, a little
silver inflation might return the economy to an 1890 level, but
that condition itself was sufficiently unjust to have generated the
reform movement in the first place. Most reassuring of all to
veteran Populists was the fact that Democrats could not go
beyond free silver to something of more substance without
toppling over into the People's Party. If they did they would be
welcomed, of course, for on that basis political realignment
would have been structurally achieved in the name of authentic
monetary reform. The silver solution could not save the people,

and that fact would soon become evident. When it did, the People's Party would be in a position to inaugurate the real solutions of the Omaha Platform.

3

The contesting positions of the advocates of Populism and those of silver fusion manifestly were not compromisable. One reflected the political purposes that had created the third party; the other reflected the political adversities the party had encountered. It is thus not surprising that the debate began on a high level of contention and rapidly became even more acrimonious. The party's national chairman, H. E. Taubeneck of Illinois, fired the first salvo after the 1894 elections when he issued a public statement that the third party had "outgrown many of the 'isms' that [had] characterized its birth and early growth" and would "take a stand on the financial question" to attract all those repelled by the "wild theories" of the Omaha Platform. This pronouncement, coming from out of the blue, stunned reform editors. Warnings flashed angrily in the editorial columns of the third party press throughout the nation. No reform editor was more strategically placed to observe the internal politics of the shadow movement than Thomas Byron, the editor of General Weaver's old paper in Iowa, the *Farmer's Tribune*. His warnings had remarkable precision:

> We are credibly informed that, as a step in this desired change of front, they will attempt to depose W. S. Morgan from the editorship of the Reform Press ready-print mat service, which they think in his hands smacks too much of the despised Omaha Platform and elect Dunning of the *National Watchman*, which is their special organ, in his place.
>
> We are getting very tired of these vexatious self-seekers, these mouthing men at Washington and elsewhere who have no visible means of support except scheming in questionable politics, and we would have exposed them by name and scheme long ago, had we not had a perfect confidence in their inability to do the party any harm. . . . The People's Party is now too intelligent, too determined, too large and too well self-governed for self-seeking, would-be bosses either to harm it or control it.

The confidence of the reform editors grew out of their belief that a move to emasculate the Omaha Platform had no chance of being taken seriously by the great mass of third party adherents. Taubeneck's apparent belief that Populists thought their platform consisted of "wild theories" represented a bad misreading of the third party constituency, and reform editors knew it. But as the debate moved into 1895, the editors began to lose their complacency, and Thomas Byron himself betrayed the reason with his elliptical reference to "these self-seekers . . . who have no visible means of support." With that phrase Byron revealed the detection by Populists of the arrival of a new element in American politics—the growing participation of silver-mining interests and the involvement of some Populist spokesmen. The proliferation of "bimetallic" leagues and other front groups for silver coinage and the sheer effectiveness of the silver lobby in penetrating certain leadership ranks of the People's Party became a growing concern. The Eastern silver publication, the *Silver Knight,* was edited by a Republican, Nevada's Senator Stewart, but reform editors belatedly discovered that the new business manager of the journal was J. H. Turner, a member of the national executive committee of the People's Party. The spectacle of third party functionaries on the public payroll of the silver lobby was unsettling, to say the least.

Soon after the 1894 elections the attention of the mine-owners focused upon the disrupted ranks of the Democratic Party. The American Bimetallic League began sponsoring scores of "conferences" at which little business was carried on beyond mass celebrations of the virtues of silver coinage. Democratic politicians found it surprisingly easy to attend these conferences. Senators, Congressmen, and would-be gubernatorial candidates flocked to meetings from Atlanta to Salt Lake City. In June 1895 an enormous spectacle was held at Memphis as a South-wide demonstration of strength. Ben Tillman showed up with an entourage of earnest new "Silver Democrats" from South Carolina, and he was amply supported by scores of Congressmen, Senators, and local politicians from throughout the South. Any sagacious political observer surveying the local and regional political influence of the participants could have properly concluded that the Gold Democracy of Grover Cleveland had fallen into serious trouble throughout the Old Confederacy. Populist voting strength in 1894, more than silver money, accounted for this new trend of 1895.

Some Populist politicians were also seen on the fringes of these silver celebrations. Joining Taubeneck and Weaver in such associations were Marion Butler of North Carolina, Senators Allen of Nebraska and Kyle of South Dakota, and Congressmen Kem and McKeighan of Nebraska. A common political fact connected the mutual devotion of these political figures to silver coinage: they all owed their seats in Washington to the politics of two-party cooperation in their own states.

Populists with good greenback memories could see that James Weaver's love affair with the cause of reform had produced a very inconstant marriage. Though Weaver gave voice to radical ideas, he manifested a decided distaste for the weakness of radical third parties. Populists belatedly recalled that Weaver owed his 1880's election to Congress as a greenbacker to an early fusion arrangement with Iowa Democrats. After his own presidential defeat in 1892, Weaver played an active role in preparations for an 1893 convention of the American Bimetallic League orchestrated by lobbyist William Harvey and was rewarded by being allowed to preside over the three-day mass meeting. Weaver worked closely with General A. J. Warner who, as "president" of the American Bimetallic League, was the nation's most visible silver lobbyist. Further evidence of silver penetration into the top ranks of the People's Party came when Taubeneck himself accepted a position as a member of a special silver committee to serve as a Washington lobby for the white metal.

4

Despite Taubeneck's pronouncement against the "wild theories" of the Omaha Platform, the national chairman's cause was a desperate one. What the silverites had in big names they sorely lacked in numbers. Most Populists did not run for office, and their respect for the Omaha Platform proved most galling to

the party revisionists. Indeed, Taubeneck's thirteen-month campaign that began immediately after the 1894 elections developed
into an educational adventure, one in which he discovered the
ideological shape of the national agrarian movement of which
he was the titular head. His need for such elementary information was a curious product of his own career as a reformer—
one that matured in an organizational and ideological environment just outside the mainstream of the agrarian revolt. Perhaps
the most telling irony in all of Populism was the third party's
selection of a national chairman who had no knowledge of the
long struggle of the Farmers Alliance in the South and West.

Taubeneck had originally made contact with the farmers'
movement through the Illinois-based Farmers Mutual Benefit
Association, and he became one of three aspiring politicians
elected to the 1891 Illinois legislature with FMBA help. Through
bizarre circumstances, he promptly emerged as a hero to radicals
throughout the nation. The 1890 election in both South Dakota
and Illinois had created a minority of Republicans in both state
legislatures that were to choose United States Senators, and while
the long years of agrarian helplessness in the Illinois affiliate of
the Northwestern Farmers Alliance had produced a radical
presence of precisely three legislators in the Illinois House (all
allied with the FMBA), the South Dakota legislature was full of
agrarian reformers. Needless to say, Democrats in neither state
wanted to see a Republican elected. Gradually, an elaborate two-
state plan of cooperation was worked out. The three agrarians
in Illinois—who held the balance of power between the two
major parties—were to vote for a Gold Democrat for Senator;
in exchange, the Democrats in South Dakota would vote for an
agrarian, James Kyle. Both Democrats and Populists would get
a Senator and the Republicans would get nothing. The bargain
was carried out by the Democrats in South Dakota, sending a
surprised and overjoyed Kyle to Washington. But Taubeneck
proved recalcitrant in Illinois. To the delight of radical Alliancemen across the nation, he took the rather candid and
persuasive position that he had not run for office in order to
dispatch goldbugs to Washington. A deadlock raged for weeks
before Taubeneck's two associates buckled under the pressure,

and sent the Gold Democrat to Washington by a margin of one vote. But Taubeneck, though defeated, earned a reputation as an incorruptible reformer. These events took place during the very weeks when plans for the Cincinnati convention of 1891 were occupying the attention of Alliancemen. Taubeneck not only attended, he left the city as the provisional national chairman of the People's Party, a position confirmed by convention action in 1892. The new chairman thus had grown up neither with the Alliance movement nor with its Reform Press Association. In his fight for free silver, he learned a great deal about both.

It turned out that Taubeneck did not see the nation's monetary system through the eyes of either angry farmers or greenback theoreticians. He voted against Gold Democrats because he thought the currency should be expanded and that the easiest way to do that was through silver coinage! He thus quite easily came to an early conclusion that Populism might gain armies of recruits if it jettisoned its platform and focused on free silver. Believing that all Populists shared what looked to him to be an obvious conclusion, Taubeneck, in December 1894, sent out several hundred notices inviting Populists to St. Louis to attend "the most important meeting since the Omaha convention." The simultaneous press statement, outlining the impending abandonment of "wild theories," ensured that the naïve party chairman would enjoy a full attendance of those he invited. They came—and they ran over him. Instead of being narrowed to a single silver plank, the Omaha Platform was broadened slightly to include declarations for municipal ownership of utilities. This effort to generate a concrete Populist appeal in the cities was led by the distinguished Chicago antimonopolist, Henry Demarest Lloyd, a circumstance that gave Taubeneck a new, though still shaky, insight into the makeup of his own party. Taubeneck announced through the *National Watchman* that the People's Party was imperiled by a socialist takeover. Populists, led by reform editors, essentially had two reactions to Taubeneck's posturing: either the chairman had lost his mind or he was on the payroll of silver mineowners. Bruised, Taubeneck decided that the next time he would take steps to mobilize the full force of his own narrow base within the party while holding attendance of all other Populists to an absolute minimum. The conference in December 1894 thus became the first and last mass meeting of Populists that Herman Taubeneck ever called.

Taubeneck's disappointing conference was but one of three critical Populist meetings in the months after the 1894 elections. In February, 1895 the National Reform Press Association convened in Kansas City. The attendance of over one hundred and fifty editors from throughout the nation signaled the degree of indignation they felt at the unexpected new course of the party's national leadership. Taubeneck was without defenders at Kansas City. Seemingly oblivious to that fact, he dispatched a highly indiscreet letter instructing the editors to cease and desist in their propagation of the Omaha Platform. A threat to withhold all advertising placed in the reform press through the national office punctuated this rather imperial edict. The editors were duly outraged and showed their disdain for free silver by reprimanding Taubeneck and resolving to defend the Omaha Platform against any and all threats. The editors even went a step further, sending a delegation to Washington to confront Taubeneck. The party chairman was induced to sign a statement that the Omaha Platform could not be altered except by a national convention of duly certified delegates. For good measure, the editors extracted similar pledges from the party's silver-minded Congressmen. Taubeneck's bid had failed: the party's politicians temporarily fashioned a lowered silhouette and the chairman again retreated.

The National Farmers Alliance also met early in 1895, and it delivered a final blow to the party chairman. The order pointedly reaffirmed all elements of its broad-gauged greenback stand on monetary issues by listing them once again in first place on the agenda of Alliance "demands." Alliance leaders also signaled a possible interest in broadening the Omaha Platform by recommending, "as subjects for discussion and education," a graduated property tax and the initiative and referendum. It was clear that the National Alliance, too, stood firmly by the Omaha Platform. The silver balloon had failed to soar: the party's rank and file, and its editors, had spoken and the Omaha Platform remained intact. The People's Party had too many Populists.

As Taubeneck belatedly came to understand, he had assaulted

the greenback heritage of the People's Party. The political tenacity and patience of greenbackers proved surprising to him. Few illustrated that characteristic better than Thomas Byron of Iowa. From his embattled outpost on General Weaver's old newspaper, the *Farmer's Tribune*, Byron wrote an article on the politics of the money issue that seemed squarely aimed at bringing both Weaver and Taubeneck to their senses:

> Free silver men in the old party ranks profess to be unable to understand why the People's Party do not jump at the chance to attract their votes and thus possibly assure the new party immediate success by just dropping the rest of the Omaha Platform. . . . Their inability to understand the Populist position is due entirely to their ignorance of the money question. . . . Thoughtful men . . . not afflicted with the itch for immediate office-seeking but . . . actuated mainly by love of country . . . deem it best . . . that . . . the party do not grow too quickly, lest the people come into it faster than they can be educated . . . and the party get out of control of its friends, and reforms enacted prove in the new hands only superficial. . . . Hence we had rather not be overrun at this critical juncture by the Silver Goths from the forests of the two old parties. . . . When they can come into the movement understandingly they shall be welcomed as brothers.

Nor was Byron alone. In Kansas, where the Populist tent had begun to flap in the breeze, Stephen McLallin of *The Advocate* hammered in some greenback stakes. *The Advocate* called the one-plank idea "the height of absurdity." As Taubeneck discovered, greenback doctrines constituted a fundamental intellectual barrier to the politics of the silver crusade.

In the aftermath of these setbacks, the nature of Taubeneck's proprietary chairmanship became even more visible. He re-oiled the ideological weaponry he had first tested the preceding December and once again declared that the People's Party was in danger of a socialist takeover. Citing the actions of the reform editors in Kansas City in support of his charge, he included as additional evidence the barrage of editorials in behalf of the Omaha Platform that had appeared in leading Populist journals. He then publicly called the roll of the offenders who had "gone over to the Socialists"—the leading journals of the agrarian revolt. The reply of the *Nonconformist* was typical of the responses evoked by Taubeneck's ukase:

> Everybody who reads the *Nonconformist* knows very well that it has never swerved an inch from the Omaha Platform or the principles of the Populist Party. . . . We insist that all the misunderstanding arose from the chairman's effort to assume an authority which did not belong to him. . . . [T]he Populist newspapers have always done their duty and have never rendered more substantial service to the party than by refusing to allow the platform to be emasculated and a side show set up in its place at the dictation of a small coterie who imagine they know it all because they have spent a few months in Washington. . . . If supporting the Omaha platform makes a man a Socialist, we are all Socialists. If it was not Socialistic to support the platform for two years up to last December it was not Socialistic to support it after that date.

In the *Farmer's Tribune*, Byron ceased his defense of the party platform and went over to the attack. He informed his readers that he saw more sinister forces at work—"silver barons and certain idle politicians." Worse, said Byron, while the reform press defended the platform, Taubeneck and Weaver were lobbying where it really counted—among the party leaders who would be organizing delegations to the Populist national convention of 1896. The Michigan state convention was said to have delcared for the new one-plank policy only after "General Weaver traveled to the state for that especial purpose." Weaver's conduct, said Byron, was "animated by an overwhelming desire to be the fusion candidate of the old party silverites and the People's Party for president in '96."

Taubeneck and Weaver were not without defenses against such attacks. Weaver persuaded the principal owner of the *Farmer's Tribune* to take editorial control out of Byron's hands. Rather than submit Byron resigned in May 1895. And in Lincoln, Nebraska, the *Wealth Makers*, another outspoken critic of William Jennings Bryan and William Allen, failed to survive the hostile climate of the shadow movement and ceased publication early in 1896.

On the whole, however, reform editors reflected the perspectives of their readers. Manning and Whitehead in Alabama,

Burkitt in Mississippi, Watson in Georgia, Norton in Illinois, McLallin in Kansas, and Park in Texas were only among the most visible of hundreds of reform editors who closed ranks to defend the party's basic policy through the period of increasing silver agitation in 1895–96.

Meanwhile, Taubeneck's cry of "socialism" was blunted by the socialists themselves. In the late spring of 1895 the Chicago coalition of Populists and Socialists* completely fell apart and suffered a crushing defeat in off-year municipal elections. Though the *National Watchman* chortled in triumph, the matter had become purely academic. The real battle to define Populism remained in the larger national arena that involved the monetary politics of all three parties and the respective instruments of persuasion for Republican gold, Democratic silver, and Populist greenbacks.

5

The lobbying laurels, it turned out, were won by the silverites. In the winter of 1894–95 they struck a bonanza when the third pamphlet published by William H. Harvey out of the Chicago office of the American Bimetallic League became a national best seller. "Coin" Harvey had produced *Coin's Financial School*. At twenty-five cents a copy, this cleverly written and profusely illustrated pamphlet of 155 pages enjoyed a surprisingly large sale. As soon as they discovered they had a hit, silver lobbyists purchased and distributed hundreds of thousands of copies of the book in numerous and varied editions. Millions suffering from the maladies of depression and a contracted currency read Harvey's appealing, if wildly inaccurate, analysis of the monetary system and of various gold conspiracies by international bankers. "Coin" Harvey not only recreated much of the atmosphere of hysteria and conspiracy that had accompanied the furor over the "Crime of '73" a generation earlier,† he even paraphrased

---

* An extended account of the coalition of Populists and Socialists in Chicago is available in Chester McArthur Destler, *American Radicalism*, and a shorter summary is contained in Goodwyn, *Democratic Promise*, pp. 411–21.

† See Chapter 1.

the very slogans put forward in support of the first silver drive. (The goldbug response—again recapitulating the debates in the 1870's over redemption—were scarcely on a higher intellectual plane; the argument that "money was only as good as the gold that is in it" was just as erroneous in 1895 as it had been after the Civil War.) But slogans, particularly when enhanced by the conspiratorial reasoning of "Coin" Harvey, were, for the time being, sufficient for many thousands of the Democratic faithful, hard-pressed as they were by the deepening depression. By April 1895 a Mississippi Congressman was writing to one of Cleveland's cabinet members of the sudden fame of the new silver propaganda book: "A little free silver book called "Coin's Financial School" is being sold on every railroad train by the newsboys and at every cigar store. . . . It is being read by almost everybody." Kenesaw Landis reported the same phenomenon from Illinois: "The God's truth is the Democratic Party in Indiana and Illinois is wildly insane on this subject. . . . The farmers are especially unruly . . . utterly wild on the money question. You can't do anything with them—just got to let them go."

The struggle for the soul of the People's Party increased in intensity throughout 1895 as Populist politicians seeking coalition with Democrats wavered, reaffirmed the Omaha Platform, then appeared to embrace the silver solution.

Weaver's elusiveness stirred deep doubts in Davis Waite, the leader of Colorado Populism. "I hardly know what to think of Weaver," he wrote in confidence to Ignatius Donnelly. Weaver was "in sympathy with Bland and Bryan . . . a Democratic movement." Weaver had assured Waite that he had lost confidence in the American Bimetallic League and was prepared to stand firm on the Omaha Platform, but, said Waite, "He came into Colorado in the pay of that outfit and undertook to stampede the Colorado Populists on the single issue of silver."

Others were less disingenuous. An increasing number of Kansas politicians said they wanted the Omaha Platform, but they wanted victory, too. The simple fact of the matter was that the silver lobby was making silver popular.

6

After twelve months of active but frustrating campaigning for "a union of reform forces," Weaver, Taubeneck, and their associates decided at year's end to take authoritarian steps to control the excess of democracy in the People's Party. On the final day of 1895, Weaver confided the results of fusionist planning to William Jennings Bryan. "We have had quite enough middle of the road nonsense, and some of us at least think it about time for the exhibition of a little synthetic force." The first "exhibition" of the new policy came with considerable clarity two weeks later. At a mid-January meeting Taubeneck and Weaver implemented a shameless plan to stack the Populist national convention. After a year of cajolery and hints of imminent silver campaign money, Taubeneck had made head-way in diminishing the presumed importance of the Omaha Platform to many members of the Populist national committee. Taubeneck's increased influence on the committee came principally from states having little or no Populist presence—New England; the mid-Atlantic states, parts of the upper Midwest and border states. Adding these to the mining regions of the West that had always been preoccupied with silver, Taubeneck had acquired what he believed to be a majority for silver. He was able to get approval for a surprising plan—an apportionment of delegates to the Populist national convention that markedly under-represented the South. In contrast, the industrial East and selected states with known pro-silver leaders were to be substantially over-represented. The reform press, predictably, raged about such boss tactics, and mid-road forces promised to bring the entire matter up for review at the national convention itself.

But whatever strategy Taubeneck presumed he and his colleagues were following, the silver mineowners—and William Jennings Bryan—were clearly proceeding toward another objective. Their plan manifestly called for capturing the Democratic Party so the "reform forces" could unite under that somewhat broader umbrella. After nominating a silver ticket, the Democrats would, in effect, confront the Populists with the choice of supporting the major party or ignoring the claims of its own fusionist wing to be working to "harmonize all reformers."

Under these circumstances, mid-roaders concluded that the raw practical politics of silver promised to bring unity in a way that would destroy the People's Party. The way to avoid that fate was to hold the Populist convention first, nominate a sound third party reformer for President, and leave to the Democrats any onus attached to dividing the "reform" forces—granting that such a multi-party entity existed in the first place. While the advice of the *Southern Mercury* to this effect was brusquely ignored by Taubeneck, the implementation of an alternate strategy required delicate coordination with both Democrats and the silver lobby. A flurry of correspondence between Taubeneck, Weaver, Butler, Allen, Donnelly, Bryan, and silver lobbyists revealed that the Democratic National Committee planned to meet on January 16, 1896, to select a date for the Democratic convention. Taubeneck thereupon called his own national committee together on January 17, and Weaver privately assured Bryan that the People's Party committee not only would vote to meet after the Democratic convention, but that Populist leaders would even postpone the decision on time and place until they could consult with the chief silver lobbyists.

When the Democrats selected July 7 in Chicago for their convention, Taubeneck affirmed his control over the Populist national committee—and the committee's subservience to the silver strategists—by postponing for a week the decision on the date of the Populist convention. With plenary powers to act for the People's Party, the fusion leaders promptly entrained for Washington to meet with the lobbyists of the American Bimetallic League. With the silverites' permission, or acquiescence, as the case may be, Taubeneck announced from Washington on January 24 that the Populist national convention would be held on July 22 in St. Louis, two weeks *after* the Democratic convention. It was as if every move were calculated to prove that the People's Party itself had become superfluous.

Thereafter, the silverite strategy unfolded in a way that must have brought great satisfaction to the American Bimetallic League and to a new organization called the "Democratic

National Bimetallic Committee," which suddenly materialized with an adequate treasury. The cause of silver became a visible force in the spring months of 1896, as silverites won control of the Democratic party in state after state across the South and West. The functionaries of the silver lobby kept the Populist fusion leaders off balance during the entire period by alternately providing assurances that goldbugs would control both major parties and by dangling silver money to finance "the approaching campaign." Gerrymandered and slandered by their own party's leadership, Populists watched these developments with growing alarm. The "little synthetic force" that James Baird Weaver had suggested to squelch the Omaha Platform had indeed materialized. Through "Coin" Harvey, the cause of silver was penetrating Democratic ranks from below; through Herman Taubeneck and his small circle of friends, it was penetrating the ranks of the People's Party from above. Silver mineowners directly financed both efforts.

7

The aspirations that had produced the agrarian revolt proved too strong, however, to subside passively in the face of the campaign of the silver lobby. A driving energy persisted in the People's Party, one that merged ideological assertion with a remarkable degree of collective purpose. This circumstance was traceable to the existence of shared memories that, in 1896, stretched back over almost a full generation of hope and effort. And no one, it turned out, was more committed to the Omaha Platform than the Alliance founders in Texas.

The radicalism of the Texas third party often achieved an intensity that in retrospect is not easy to describe. It combined the soft admonitions of Judge Thomas Nugent, the skillful greenback arguments of J. M. Perdue and Harry Tracy, the tactical radicalism of William Lamb and Evan Jones, and an astonishing degree of rank-and-file militance on the part of farmer-veterans of the Alliance movement. When the American Bimetallic League first suggested the possible need for a new national silver party, Nugent wrote a reply that was characteristically quiet in tone and radical in content. Silver coinage, he said, would "leave undisturbed all the conditions which give rise to the undue concentration of wealth. The so-called silver party may prove a veritable trojan horse if we are not careful." Other Texans expressed similar views with considerably less reserve. As the *Mercury* viewed matters, "With Debs in jail, no American is a freeman." Full-scale editorials produced phrases about the "bitter and irrepressible conflict between the capitalist and the laborer," a circumstance requiring "every wage earner to combine and march shoulder to shoulder to the ballot box and by their suffrage overthrow the capitalistic class." These were special moments in the *Mercury*'s editorial career, however. More often the paper controlled its radical ardor sufficiently to combine humor with Populism, as when it catalogued the plethora of Populist candidates who wanted fusion and concluded that the "free silver party is like a Kentucky militia regiment. It consists of all colonels and no privates."

As the most authoritative voice of the Reform Press Association in the South, the *Mercury* possessed the credentials and its editor the temperament for a showdown encounter with the fusionists. The *Mercury* counseled, prodded, praised, admonished—extending advice to Taubeneck in 1894, to Simpson and Weaver in 1895, and to Butler in 1896. Radical journals like the *American Nonconformist* and the *People's Party Paper* were praised, while Dunning's *National Watchman* was roundly condemned. The labor movement received a mixed review. Eugene Debs's *Railway Times* was "an earnest, fearless, and honest friend to labor," and the Knights of Labor journal in Chicago was similarly blessed for being "beyond the reach of politicians and the money gang." But Samuel Gompers and much of the rest of labor got low marks in steadfastness, a fact which kept "organized labor disorganized." The Kansas brand of Populism, intermittently beset by fusionist tendencies since 1892 (though usually on the basis of the Omaha Platform), had lost the *Mercury*'s approval quite early, and the paper fairly chortled in triumph when the 1894 returns showed the Texas party, all radical banners unfurled, had outpolled the Kansans in the November elections. As for the Nebraska fusionists, the *Mercury* could scarcely conceal its contempt.

But no single Populist had a longer and more intense memory
of the agrarian revolt than the pioneer "Traveling Lecturer"
who had carried the order's cooperative message from the
original frontier counties in Texas back in the winter of 1883–84.
If Herman Taubeneck had come to the People's Party from
outside the Alliance experience, S. O. Daws of Parker County,
Texas, came as close as any man to being the symbol of that
experience. When Daws heard of Taubeneck's plan to convene
a joint meeting of Populist politicians and the national committee
to set a late convention date, his indignation at this slur on "the
plain people" knew no bounds:

> Why did not the populist national committee invite the Farmers
> Alliance to meet with them at St. Louis? Why did they not
> invite other industrial organizations? Don't deceive yourselves
> gentlemen. . . . The farmers will be on hand at St. Louis to give
> their views in no uncertain sound. . . . The laboring people
> are tired of modern party politics. . . . Their diagnosis of the
> case may seem absurd to modern demagogues, and their
> remedies may seem visionary, but they are honest and des-
> perately in earnest. We propose to stand by the demands of
> the industrial people!

Six weeks before the convention the *Mercury* sifted the Populist
leadership ranks and selected its presidential candidate. "The
way to discourage these trimmers and wreckers is to place in
nomination candidates of whom there is not a shadow of a doubt
as to their honesty of purpose and adherence to the people's
cause, candidates who can stand squarely upon the Omaha
Platform without spending their time hiding in some obscure
corner of it." The *Mercury* advised the various state parties to
send delegates to St. Louis "who are honest and sincere, pledging
them to support the Omaha Platform in its entirety, and
instructing them to vote for the most broadminded statesman
and patriot of the century, Eugene V. Debs, for president."

The mid-road crusade activated all the resources of the reform
movement that were free of the influence of Herman Taubeneck
and the national silver lobby. Across the nation, these resources
were imposing. In employing them, Populists made a conscious
effort to strew in the path of the silverites as many practical and

symbolic roadblocks as possible. The Reform Press Association, for example, delivered the clearest message it could to the fusionists by electing Milton Park as its national president in 1895. In 1896, the association renewed its long support for W. Scott Morgan's Populist ready-print service and mobilized its internal machinery to disseminate to all reform weeklies throughout the nation the necessary facts, figures, and arguments in support of the party's platform. In the same weeks the National Farmers Alliance issued a ringing challenge to the "platform wreckers" in the form of a new preamble to the famous Alliance demands. The statement climaxed the insurgent heritage of the Farmers Alliance with the most radical declaration of its entire history:

> We hold therefore that to restore and preserve these rights under a republican form of government, private monopoly of public necessities for speculative purposes, whether of the means of production, distribution or exchange, should be prohibited, and whenever any such public necessity or utility becomes a monopoly in private hands, the people of the municipality, state or nation, as the case may be, shall appropriate the same by right of eminent domain, paying a just value therefor, and operate them for, and in the interest of, the whole people.

With this new peg anchoring their left flank, the defenders of the Omaha Platform entered a final pre-convention phase of practical politics. The chairman of the Texas People's Party, H. S. P. "Stump" Ashby, the "famous agitator and humorist," embarked on a convention-organizing tour through the South. One of the most effective orators of the agrarian revolt—and certainly one of the most experienced—Ashby moved to offset Taubeneck's gerrymandering scheme by cementing the Southern convention delegations firmly to "the middle of the road."

As the time approached for the various party conventions, silver advocates in the Democratic Party consolidated their hold on state delegations throughout the South and West. But within the People's Party the tactical position of silverites began to deteriorate, and the mid-road campaign began to bear fruit.

The steady pounding by Alliance-greenbackers and reform editors produced sudden cracks in supposedly solid fusionist state delegations. Taubeneck reported gloomily to Donnelly that "something is out of joint in Indiana," and a Missouri Populist fired off a warning to Marion Butler that silver was a trap set by Democrats who "simply run with the stampeding cattle until they circle them back into camp." Alliance leaders occupying prominent positions in the California People's Party had long defended the Omaha Platform against silver politicians, and they stepped up their criticism of fusion tactics as the convention approached. Willits of the Kansas Alliance, supported by such party luminaries as William Peffer, Annie Diggs, and G. C. Clemens, did the same. Mississippi and Arkansas rallied to the Omaha Platform, and Joseph Manning in Alabama defended the Populist cause against the decamping Reuben Kolb. The latter, having never quite made the break with "the party of the fathers," had a shorter return distance than most other Populist leaders throughout the South. The Arkansas third party, well-honed for years by the educational program of the Wheel, the Alliance, and W. Scott Morgan, was united against the silver panacea in all its forms. Though the Alliance in Nebraska had long since deteriorated, a handful of radicals from the Kearney County cooperative movement manfully raised the Populist flag in the very citadel of Democratic-Populist coalition. On the eve of the 1896 Populist national convention the leader of the Georgia delegation, raised on years of Watsonian teachings, wired that the Georgians were coming to St. Louis "in the middle of the road." Similarly, the old editorial flagship of the North Carolina Alliance, L. L. Polk's *Progressive Farmer*, warned the Populist faithful that the fusionists were attempting "to deliver the entire People's Party into the lap of the Wall Street Democracy at one time."

A pattern was clear: where the roots of the agrarian revolt sank deepest the Populist dream of broad-based reform had its strongest support. Old Alliancemen were, indeed, "heavyweight Populists" who, along with one of the oldest of them all, S. O. Daws, proposed "to stand by the demands of the industrial people." By June 1896, some of them had been at the work of reform for almost twenty years. They went to St. Louis in anxiety, but not without some confidence. After all, Populism was, as they told themselves, their movement.

8

Yet while Populists successfully mobilized both the reform press and the Farmers Alliance in a successful defense of the Omaha Platform—an ideological victory won wholly within the ranks of the People's Party—the silverites won the larger battle for the attention of masses of Democratic Party voters in the South and West. For while the Republicans struck a blow for traditionalism by nominating William McKinley on a gold standard platform, the agrarian revolt had shaken the very foundations of the post-Civil War Democracy. Southern and Western Democrats, reeling from six years of the Populist onslaught, were determined to refurbish the party's appearance by writing a silver platform and nominating any one of a half-dozen silver politicians as its presidential candidate.

As the advance guard of Democrats began arriving in Chicago for their convention, Taubeneck at last perceived the new balance of forces within the old party and the political myopia of his own position. He immediately abandoned his long-announced objective of "uniting the reform forces of the nation." In a desperate effort to stem the silver tide, he warned the Democrats that the People's Party would nominate its own ticket even if the Democrats should select a candidate friendly to silver. The Populist chairman thereupon discovered how his years of propagandizing for silver had weakened the third party's political leverage. The Democrats—and their friends in the silver lobby—simply ignored him. Both the Democrats and the silver lobby had looked after their own interests; if Taubeneck had not, the problem was his, not theirs.

The Democratic convention confirmed Taubeneck's worst fears. Though Missouri's "Silver Dick" Bland was the early favorite to win the nomination, William Jennings Bryan's "Cross of Gold" speech transported the delegates. Bland stepped aside and eased the way for the nomination of Bryan. As if to emphasize their Democratic orthodoxy, even at the moment of apparent party schism, the delegates, including whole state delegations pledged to Bryan, supported the vice presidential candidacy of Arthur Sewall, a conservative banker and shipping magnate from Maine. In both of its standard bearers the silver crusade thus possessed spokesmen who were intellectually and politically committed to a hard-money, redeemable currency. Greenback principles were anathema to the new crusade—as they had been during the first silver agitation over "the Crime of '73," a generation earlier.

Dedicated as Herman Taubeneck was to the cause of silver, the truly distressing product of the Democratic convention lay in another political reality: the People's Party, he now saw, was not only trapped, it was threatened with the loss of its own identity. His strategy in ruins, the distraught fusionist leader fell silent. He made his way to the Populist convention city of St. Louis immediately following the Chicago debacle and "sullenly refused to be interviewed."

The hyperactive days of Herman Taubeneck had come to an end. He would not head a grand "union of forces," and the thought stripped him of his political will. On the eve of the third party's most important convention, the Populist national chairman declared his "neutrality" and receded from view.

The men whom Herman Taubeneck had rallied to his fusionist cause could not, however, afford such a passive response. They were politicians all—and politicians who had to find a way to put the best possible face on embarrassing events. Jerry Simpson, out of office for two years, easily swallowed the Democratic medicine: "We should adopt our own platform and nominate Bryan. That would unite the Democrats and Populists on the silver issue." In a vague straddle, Simpson said he favored the Omaha Platform "with a few elisions." With considerably less strain, so did almost the whole of the Nebraska third party which had been created in cooperation with Bryan. Ignatius Donnelly, the old greenbacker, also appeared to be wavering. "Narrow Populism to free-silver alone," he wrote in 1895, "and it will disappear in a rat hole." But after the Democratic convention the Sage of Nininger wrote in his diary: "Exciting times these.

Shall we or shall we not endorse Bryan, but I do not feel that we can safely accept the Democratic candidate. I fear it will be the end of our party."

A sober political reality underlay this lack of fixed purpose on the part of so many Populist politicians in the West. All owed their seats to the joint support of Populists and Democrats in their own states. If the People's Party nominated a Populist ticket against both Bryan and McKinley, Populist officeholders in the West were doomed to defeat by a division of their electorate, and they knew it.

9

Populists thus arrived in St. Louis for their convention to discover that the fusion strategists, far from retreating, had advanced to new ground: the third party should nominate the Democratic ticket of Bryan and Arthur Sewall. To block this strategy—which frankly appalled them—mid-roaders called a mass meeting of delegates at the Texas headquarters in the Southern Hotel in St. Louis to map plans to retain the substance of the Omaha Platform. Delegates from twenty-three states attended the mid-road caucus. Southern third party men circulated among the delegates and indignantly reviewed the fraudulent practices, wholesale intimidation, and election-day violence that had characterized the conduct of the Democratic Party in the South. How could reformers in the North and West ask their Southern comrades to embrace their most hated foes? "Free silver" did not purge the Southern Democracy of its inherited style. The message was clear: fusion with the Democratic Party would destroy the morale of Southern Populism.

Early in the proceedings, during balloting over credential contests between rival delegations in various states, the strength of mid-road sentiment became evident. The contests culminated in a successful campaign led by "Stump" Ashby of Texas to seat as part of the Illinois delegation a Chicago socialist contingent affiliated with Eugene Debs and Henry Demarest Lloyd. On such a highly controversial issue as a public endorsement of socialists, the margin of the convention's decision was narrow, 665 to 642, but the mid-roaders were jubilant.

They suffered from a critical shortcoming, however—their continuing absence of "big names," essentially traceable to Democratic frauds that had deprived Populist candidates in the South of election. This circumstance became crucial when the convention made ready to fill the important position of permanent chairman. The only mid-road Populist of genuine national reputation was Georgia's Tom Watson, who had originally been elected in 1890 before vote frauds became a way of life in the Gilded Age South. But Watson was not in attendance in St. Louis. The fusionists, on the other hand, could call on most of the party's officeholders in the West. They selected Nebraska's William Allen. The mid-roaders finally settled upon an obscure Maine radical named James E. Campion. Allen won by a vote of 758 to 564. It was no minor defeat for the mid-roaders: with the election of Allen, any part of the nominating process that turned on rulings from the chair passed firmly into the hands of the fusionists. As they had enjoyed throughout Taubeneck's long tenure, fusionists had the benefit of a commanding tactical position, augmented by their earlier moves to gerrymander and stack the convention.

Nevertheless, the fusionists were asking a great deal—not only that the delegates water down their basic platform and accept Bryan, but also that they accept as their vice presidential nominee a financier and businessman with known anti-labor proclivities. Sewall's nomination constituted a positive repudiation of virtually the entire political history of Populism. Mid-road efforts were not wasted: in a definite blow to the fusion agenda, the convention voted 738 to 638 to nominate a vice presidential candidate first—a move that mid-roaders confidently expected would place a Populist on the ticket and make it impossible for Bryan to accept the Populist nomination. Despite herculean efforts, the fusionists also lost the battle on the platform. The task of shepherding the platform committee through a wholesale ideological retreat from the party's basic principles was entrusted to James Weaver. Though he had some minor successes, he failed on all the principal issues. While Weaver sidetracked mid-road attempts to broaden the platform by adopting woman suffrage, the recommendations of the Farmers Alliance for direct legislation by initiative and referendum were approved. Beyond this, the need for government ownership of the means of transportation and communication, the land planks and, most important of all, the essential monetary planks of the greenback heritage were all reaffirmed. Thus, on the great triumvirate of Populist issues—land, finance, and transportation—the tenets of the Omaha Platform were reincorporated in the new "St. Louis Platform."

Though newspaper reporters were at a loss to locate the prevailing balance of power in the Populist convention, there was a "sense of the convention." On issues where fusionist "big names" were placed in nomination against unknown mid-roaders fusionists prevailed, as in the election of Allen; where fusionist "big names" contested basic greenback tenets of Populism the fusionists were defeated, as when Weaver lost the platform fight; when the candidacy of Bryan was counterpointed against the preservation of the People's Party, the fusionists again lost, as in the decision to nominate the vice presidential candidate first. But with all this, a decisive number of delegates—centered in the over-represented states of the East and upper Midwest, where the third party was weak—felt that the various forces in the nation arrayed against "the gold power" should not be "divided." Since the Democrats had named Bryan, clearly an "anti-gold" candidate, they felt the Populists should follow suit—as long as such an act would not destroy the People's Party. The difficulty of the fusionist position lay in this latter qualification, for the circumstances essential to the preservation of a Populist presence were unacceptable to the Democratic Party. This soon became clear to all.

After the Populists had decided to nominate a vice presidential candidate first, the principal Democratic representative at the convention, Senator James K. Jones of Arkansas, the party's national chairman, wired Bryan of the need for him to state his position to the convention should it select anyone but Sewall for the vice presidency. Bryan forthwith telegraphed a reply to Jones: he said that he could not accept the Populist presidential nomination unless Sewall was his running mate. Jones, overestimating fusionist strength in the convention, thereupon began showing the reply to delegates in the expectation that such information would ensure the nomination of Sewall. Nevertheless, Sewall lost heavily to Georgia's Tom Watson, the candidate of the mid-roaders. The message was clear: the People's Party did not wish to destroy itself.

Victorious on all crucial issues, the mid-roaders retired from the third day of deliberations with renewed hope. They awoke the next morning to find Bryan's reply to Jones spread across the pages of the St. Louis newspapers. The Democratic nominee was explicit: Jones was to "withdraw" his name if Sewall was not nominated. Each session of the Populist convention had been awash in rumors, but seeing Bryan's stark reply in the newspapers lifted the subject out of the category of rumor. They could not have Bryan if they did not take Sewall, and they had not taken Sewall. The delegates may have preserved their party's identity, but now, on the morning of the fourth day of their convention, they had to nominate a candidate for President. What were they to do?

10

As the delegates gathered on the final day, the convention hall filled with a new rash of rumors. New answers from Bryan were being awaited, or had been received, or would be announced shortly. General Weaver, the 1892 Populist standard-bearer and now a much embattled fusionist, arose to deliver the nominating address for Bryan. In a speech that was adroitly vague on essential issues, Weaver told the convention that he regarded Bryan's message as "a manly dispatch." "No man could have done less," he said, "and remain a man." Whatever this meant, it assuaged few of the delegates' qualms, though it did serve the tactical function of preventing a complete collapse of the fusion plan. Bryan could not be prevented from receiving the nomination—if only his name could logically be *placed* in nomination. The politics of the convention came to hinge on this point. Bryan's telegram in effect meant tactical victory for the mid-roaders; Allen, as convention chairman, attempted to outflank the mid-roaders by blandly denying he had received it.

The delegates grew increasingly restless, and mid-roaders

from a number of state delegations made their way across the
convention hall to gather around a large mid-road banner in
the midst of the Texas delegation. For tense moments, the two
wings of the party confronted one another physically. Fights
broke out. "It was a howling mob," decided the *Post-Dispatch*.
The chairman of the Texas delegation, Ashby, somehow got to
the convention podium and, in a pointed sally, announced that
"Texas was ready to endorse Bryan if Bryan would endorse the
platform adopted." The remark brought forth a new surge of
contention and applause. The suggestion was "not well received"
by fusionists, who were suddenly worried by the volatile mood
of the convention. The nomination of Bryan would be assured
if the convention could somehow be kept under control, but
that prospect seemed increasingly uncertain. The roll call had
to be speeded up and the proceedings brought to a close.
Chairman Allen recognized Weaver's 1892 vice presidential
running mate, James Field of Virginia, for a motion from the
floor. In a step that caught mid-roaders by surprise, Field moved
that the rules be suspended so that nominations could cease.
The action had the merit of providing a means of overcoming
the excess of democracy in the Populist convention by permitting
Bryan's nomination to be achieved not only speedily, but by
voice vote. Allen quickly called for a voice vote on the Field
proposal and declared that the motion to suspend the rules had
carried.

The convention then erupted in pandemonium. Milford
Howard, the Populist Congressman from Alabama, forced his
way through protecting fusionists, creating a wild disturbance
in the rear of the rostrum, even as the entire 103-man Texas
delegation stormed to the very edge of Allen's podium. The
Nebraskan hesitated, aware, perhaps, that a raw power play
from the chair might endanger its occupant. He decided to let
the roll call continue, but when Colorado's name was called the
chairman of that state's delegation suddenly cast forty-five votes
for the Field motion to nominate Bryan by acclamation.
Hundreds of delegates stood on chairs and demanded recognition as Allen, gaveling for order, magisterially prepared once
again for a voice vote that would settle the matter once and for

all. A chorus of outrage from delegates halted this move, and Allen then announced that when the roll was called, delegates could vote for Bryan or for anyone else. This ruling provoked another storm of protest and, in the words of a fascinated reporter, "only confounded the confusion." Fights again broke out, and the convention hovered on the edge of a riot. Finally, "as the only way of restoring order," General Field withdrew his motion, and Allen permitted the roll call of states to proceed. Populists were to be permitted to nominate a candidate.

Ironically, Chairman Allen's shaky position was almost demolished by Democratic representatives of William Jennings Bryan. Bryan's message to the convention had been addressed not only to Democrat Jones but also to Populist Allen. Two other fusionist representatives, Democratic Governor Stone of Missouri and Thomas Patterson of Colorado, knew that Allen had Bryan's telegram in his possession at the moment the Populist Senator had made his first denials to the mid-roaders. Apparently feeling that their own personal ethics were impeached by Allen's conduct, Stone and Patterson implored the Populist Senator to be candid to his own delegates. Allen refused, for the plain reason that candor would have been fatal to his cause. When the Democratic onlookers finally insisted on disclosure, Allen continued his public denials and privately took refuge in demagoguery: "the Populist convention," he said to them, "will not to be run by Democrats!" That Allen's own purpose was to implement Democratic fusion strategy, while Patterson and Stone also wished to do so, but on a somewhat more honorable basis, underscored Allen's ethical bankruptcy. As a man who had once confided that "a political party has no charms for me," the spokesman for the shadow movement in Nebraska thus made explicit his determination to deceive the People's Party into a nomination of Bryan, even at the cost of destroying the party's separate identity in American politics.

In a last desperate expedient, fusionists attempted to calm delegate fears about the party's loss of autonomy by circulating rumors that Sewall had agreed to remove himself from the joint Populist-Democratic ticket so that Georgia's Tom Watson could be Bryan's running mate. The departure of Sewall would have preserved a measure of the structural integrity of the third party organization, and the rumor had great influence with wavering delegates. But it was not true.

The mid-roaders, unable to find a "name" candidate after Eugene Debs decided to decline their presidential offer and Ignatius Donnelly, wavering to the last instant, was unable to win the full support of his own Minnesota delegation, settled upon an old time Chicago greenbacker and reform editor named S. F. Norton. The relatively obscure Norton stirred little enthusiasm, even among those delegates desperately searching for a solution that was both honorable and practical. The Texans, now frantic in their effort to puncture one final and decisive hole in a leaking ship of fusion that appeared to be making port while sinking, renewed their public interrogation of Allen. Three times during the long roll call that followed, Ashby—his own voice augmented by the organized shouts of his delegation— halted the balloting by formally inquiring of the convention chairman if he had received a communication from Bryan. Each time, Allen denied that he had. Many delegates did not believe him. Others, anxious for days, remained deeply uneasy. No one wanted to destroy the People's Party. The roll call droned on toward its conclusion in an atmosphere muted by irresolution. Finally, Senator Allen announced that the People's Party had nominated the Democratic Party's standard-bearer, William Jennings Bryan, as its own candidate for President.*

It no longer made any difference what Bryan thought of the Populist platform; it no longer mattered whether he would accept Tom Watson as a running mate. He was nominated. In a convention wracked by chaos and haltingly stabilized only by the disingenuous statements of the convention chairman rep-

---

* Though many in their ranks bowed to the inevitable at the end, some 200 diehard mid-roaders from other states joined the 103 Texans in voting against Bryan. The final tally was Bryan 1042, Norton 321. This result has been interpreted by some historians as a valid indication of comparative fusion and mid-road strength in the People's Party. Indeed, the presumption that the shadow movement of free silver was the essence of "Populism" may well have its origin in this simple statistic! To achieve this conclusion, however, it is necessary to ignore the history of the agrarian revolt from its inception in 1877 through the Democratic convention of 1896.

resenting the shadow movement in Nebraska, the strategy of the silverites had prevailed.

The convention was quickly adjourned. Knots of mid-roaders gathered in an effort to discover a course of action that would save the Populist cause. They discussed the few options remaining open to them and made a few desperate plans. Then, disheartened and defeated, they left the hall.

The democratic agenda embedded in the Omaha Platform had shrunk to the candidacy of a Democrat named Bryan. The cause of free silver was intact. The agrarian revolt was over.

9
The Irony of Populism

"The People" vs. "The Progressive Society"

The foundations of modern America were constructed out of the cultural materials fashioned in the Gilded Age. The economic, political, and moral authority that "concentrated capital" was able to mobilize in 1896 generated a cultural momentum that gathered in intensity until it created new political guidelines for the entire society in twentieth-century America. Not only was previously unconsolidated high ground captured in behalf of the temporary needs of the election of 1896, but the cultural tactics tested and polished during the course of the campaign for "honest money" set in place patterns of political conduct that proved to be enduring. After McKinley's impressive victory in 1896, these patterns became fully consolidated within the next generation of the Progressive era and proved adequate during a brief time of further testing during the New Deal. They have remained substantially unquestioned since, and broadly describe the limits of national politics in the second half of the twentieth century. The third party movement of the Populists became, within mainstream politics, the last substantial effort at structural alteration of hierarchical economic forms in modern America.* Accordingly, twentieth-century American

---

*The point, here, of course, is that the liberal and socialist alternatives discussed in this concluding chapter were, respectively, either not substantive or were culturally isolated and outside the mainstream of American political dialogue.

reform has in a great many ways proven to be tangential to matters the Populists considered the essence of politics. This reality points to the continuing cultural power exerted by the political and economic values which prevailed in the Gilded Age and which today serve to rationalize contemporary life and politics to modern Americans.

The narrowed boundaries of modern politics that date from the 1896 campaign encircle such influential areas of American life as the relationship of corporate power to citizen power, the political language legitimized to define and settle public issues within a mass society yoked to privately owned mass communications and to privately financed elections, and even the style through which the reality of the American experience—the culture itself—is conveyed to each new generation in the public and private school systems of the nation. In the aggregate, these boundaries outline a clear retreat from the democratic vistas of either the eighteenth-century Jeffersonians or the nineteenth-century Populists.

2

Understandably, during such a moment of cultural consolidation priorities were not quickly isolated or identified; it took awhile for the full implications of the era to become evident. But the power of the hegemony achieved in 1896 was perhaps most clearly illustrated through the banishment of the one clear issue that animated Populism throughout its history—the greenback critique of American finance capitalism. The "money question" passed out of American politics essentially through self-censorship. This result, quite simply, was a product of cultural intimidation. In its broader implications, however, the silencing of debate about "concentrated capital" betrayed a fatal loss of nerve on the part of those Americans who, during Populism, dared to speak in the name of authentic democracy. Since the implications were so huge, a brief recital of some relevant specific details seems in order.

The enormous success of *Coin's Financial School* induced gold-bugs to counterattack in 1895–96 through the writings of a

University of Chicago economist named J. Laurence Laughlin. Laughlin produced not only theoretical works but also books, articles, and pamphlets for popular consumption. His widely syndicated newspaper column imparted an aura of scholarly prestige to the sound money cause, though his journalistic efforts, like his other writings, were almost as conceptually flawed as "Coin" Harvey's efforts. Yet Laughlin's campaign in behalf of the gold standard drew no critics outside the ranks of Populism, for the nation's university faculties were solidly "gold-bug." Among respectable elements of American society, green-back doctrines were culturally inadmissible.

However, in 1896 a young Harvard economist, Willard Fisher, decided he personally had endured enough of the currency theories of both "Coin" Harvey and Professor Laughlin. One of the nation's better-informed students of monetary systems, Fisher penned a biting attack on the two competing advocates of metallic-based currencies. Entitled " 'Coin' and His Critics," Fisher's article appeared in the *Quarterly Journal of Economics* in January 1896. Fisher treated with gentle tolerance "Coin" Harvey, whose writings, while badly "flawed," nevertheless produced a number of insights that were "intelligent." He reserved his harsher adjectives for goldbugs, and particularly for his academic colleague, Professor Laughlin. The latter, among other things, was "wrong." In the aggregate, the sheer momentum of Fisher's critique of the arguments of goldbugs and silverites carried him dangerously close to an inferential endorsement of the greenback heresy. No academic *enfant terrible*, Fisher cautiously stepped around this pitfall with an oblique reference to a scholarly alternative to metallic-based currencies—one he euphemistically described as "the familiar tabular system." The reference was cordial, but ultimately noncommittal. Beyond such obscurantism the young professor dared not venture. Though a metallic currency was not an intelligent system of money, Fisher declined to say what was. The word "greenback" was avoided throughout his article. Coupled with the routine orthodoxy that ruled elsewhere in the academic world, the extreme circumspection of the Harvard economist tellingly measured the power of the cultural consolidation at that moment being fashioned in America. Certain ideas about the economy, no matter how buttressed with evidence and interpretive skill, had become dangerous.

Though the gold standard was formerly legislated into law in 1901 over scattered and desultory opposition, the financial panic of 1907 convinced the Eastern banking community of the need for a more flexible currency. J. Laurence Laughlin, having proved his mettle in 1896, received the blessings of large commercial bankers and was once again pressed into service, this time as the nation's foremost spokesman for "banking reform." Laughlin and two associates wrote the Federal Reserve Act, which was enacted into law in 1913. The measure not only centralized and rationalized the nation's financial system in ways harmonious with the preferences of the New York banking community, its method of functioning also removed the bankers themselves from the harsh glare of public view. Popular attention thenceforth was to focus upon "the Fed," not upon the actions of New York commercial bankers. The creation and subsequent development of the Federal Reserve System represented the culminating political triumph of the "sound money" crusade of the 1890's.

These developments abounded in irony. The panic of 1907 corroborated an essential feature of the analysis behind Charles Macune's sub-treasury system, for the crisis partly materialized out of the inability of a contracted currency to provide adequate capital markets during the autumn agricultural harvest. Throughout the last quarter of the nineteenth century and into the twentieth, calls on Eastern banks by Western banks for funds to move the autumn crops had created stringent shortages within the entire monetary system. While this condition worked to depress agricultural prices—and was not without its benefits to bankers in the matter of interest rates—the banking system itself broke down under these and other burdens in 1907. The demand for a more flexible currency that issued from the banking community following the panic of 1907 was oriented not to the needs of agriculture, however, but rather to the requirements of the banking community itself. Thus, while the 1912 report of the blue-ribbon National Monetary Commission recommended new legislation establishing adequate credit for the nation's farmers, the Federal Reserve Act written by Laughlin and his associates failed to follow through. Though proponents of the Federal Reserve System often described the twelve regional banks established by the act as "cooperative banks" specifically designed to meet the impasse in agricultural credit—a description particularly prominent during public discussion of the enacting legislation—they were not, in fact, so designed. The Act provided easier access to funds only for the nation's most affluent farming interests.

The Federal Reserve System worked well enough for bankers in the ensuing years, but its failure to address the underlying problems of agricultural credit became obvious to all during its first decade of operation. The severe agricultural depression of 1920–21 once again focused public attention on the problem, leading to a marginal expansion of government policies through the establishment in 1923 of federal intermediate credit banks. But the 1923 amendments effectively extended the aid only to the agricultural middle class. In no sense were the credit problems of the "whole class" touched upon in ways that Charles Macune and other Populist greenbackers would have respected. Not until the farm loan acts of the New Deal did the nation directly address the credit requirements of the family farmer. Unfortunately, unlike the Macunite plan of making direct, low-cost government loans not only to aid farmers but as a competitive pressure on bank interest rates generally, the system of New Deal government loans operated wholly through commercial banks. It thus served as an artificial prop for the prevailing financial system. In any event, by the time of the New Deal legislation, literally half the farmers in the cotton belt and the Western granary had long since been forced into landless peonage and were effectively beyond help.

A final irony was implicit in these developments—and had they lived to see it, it was one that might have proved too much for old-time greenbackers to bear. The collective effect of twentieth-century agricultural legislation—from the Federal Reserve Act of 1913 to the abrupt ending of the Farm Security Administration's land relocation program in 1943—was to assist in the centralization of American agriculture at the expense of the great mass of the nation's farmers. The process of extending credit, first to the nation's most affluent large-scale farming interests, and then in the 1920's to sectors of the agricultural middle class—while at the same time denying it to the "whole class" of Americans who worked the land—had the effect of assisting large-unit farming interests to acquire title to still more land at the expense of smallholders. Purely in terms of land-ownership patterns, "agri-business" began to emerge in rural America as early as the 1920's, not, as some have suggested, because large-scale corporate farming proved its "efficiency" in the period 1940 to 1970. In essence, "agri-business" came into existence before it even had the opportunity to prove or disprove its "efficiency." In many ways, land centralization in American agriculture was a decades-long product of farm credit policies acceptable to the American banking community. The victory won by goldbugs in the 1890's thus was consolidated by the New Deal reforms. These policies had the twin effects of sanctioning peonage and penalizing family farmers. The end result was a loss of autonomy by millions of Americans on the land.

In a gesture that was symbolic of the business-endorsed reforms of the Progressive era, William Jennings Bryan hailed the passage of the Federal Reserve Act in 1913 as a "triumph for the people." His response provided a measure of the intellectual achievements of reformers in the Progressive period. Of longer cultural significance, it also illustrated how completely the idea of "reform" had become incorporated within the new political boundaries established in Bryan's own lifetime. The reformers of the Progressive era fit snugly within these boundaries—in Bryan's case, without his even knowing it. Meanwhile, the idea of substantial democratic influence over the structure of the nation's financial system, a principle that had been the operative political objective of greenbackers, quietly passed out of American political dialogue. It has remained there ever since.

The manner in which the citizens of a democratic society become culturally intimidated, so that some matters of public discussion pass out of public discussion, is not the work of a single political moment. It did not happen all at once, nor was it part of a concerted program of repression. Martial law was not declared, no dissenting editors were exiled, and no newspapers censored. It happened to the whole society in much the same way it happened to young Willard Fisher at Harvard, silently, through a kind of acquiesence that matured into settled resignation. This sophisticated despair, grounded in the belief that hierarchical American society could, perhaps, be marginally "humanized" but could not be fundamentally democratized, became the operative premise of twentieth-century reformers. Their perspective acquired a name and, rather swiftly, a respectability always denied Populism. In 1900–1930, it was popularly recognized as "progressivism." Later, it became known as "liberalism." In such a way, a seminal feature of the democratic idea passed out of American culture. This rather fateful process was inaugurated during the climactic political contest of 1896.

3

Popularly known as the "Battle of the Standards" between gold and silver, the presidential campaign between Bryan and McKinley witnessed the unveiling for the first time in America of the broad new techniques of corporate politics. Under the driving supervision of Ohio industrialist Mark Hanna, unprecedented sums of money were raised and spent in a massive Republican campaign of coordinated political salesmanship. The nation's metropolitan newspapers, themselves in the midst of corporate centralization, rallied overwhelmingly to the defense of the gold standard. Their efforts were coordinated through a "press bureau" established by the Republican leadership. The power of the church added itself to that of the editorial room and the counting house, and the morality of "sound money" and the "nation's honor" temporarily replaced more traditional themes emanating from the nation's pulpits. Cultural intuitions about respectability, civic order, and the sanctity of commerce, augmented by large-scale campaign organizing, coordinated newspaper and publishing efforts, and refurbished memories of Civil War loyalties combined to create a kind of electoral politics never previously demonstrated on so vast a scale. Though individual pieces of this political mosaic had been well tested in previous elections, the sum of the whole constituted a new political form: aggressive corporate politics in a mass society.

The Democrats responded with something new of their own— a national barnstorming tour by the youthful and energetic Bryan. Though such undignified conduct was considered by many partisans to be a disgrace to the office of the presidency, the silver candidate spoke before enormous crowds from Minneapolis to New York City. The effort seemed merely to spur the Republican hierarchy to ever-higher plateaus of fund-raising, spending, and organization. The nation had never seen a political campaign like it, and, for one heady moment at least, Bryan thought the electorate would react heavily against such self-evident displays of political propagandizing. But Republicans were able to generate such intense feeling against the "anarchistic" teachings of William Jennings Bryan that many modest church-goers as well as industrial captains felt that no sum of money was too great to ensure the defense of the Republic from the ravages of the silverites. Indeed, one of the striking features of the 1896 campaign was the depth to which many millions of Americans came to believe that the very foundations of the capitalist system were being threatened by the "boy orator of the Platte." That was hardly the case, of course—particularly when it came to the nation's currency. A monetary system responsive to the perspectives of commercial bankers was not at issue in 1896; the relationship of the government to bankers on the matter of currency volume and interest rates was not at issue either. In view of the shared faith of both Bryan and McKinley in a redeemable currency, the entire monetary debate turned on a modest measure of hard-money inflation through silver coinage. The narrowness of the issues involved in the "Battle of the Standards" should have put strong emotional responses beyond possibility—yet the autumn air fairly bristled with apocalyptic moral terminology. Indeed, the fervor of the campaign, for both sides, was authentic: the true issues at stake went far beyond questions of currency volume, to a contest over the underlying cultural values and symbols that would govern political dialogue in the years to come.

4

A great testing was in process, centering on the relative political influence of two competing concepts—that of "the people" on the one hand and of "the progressive society" on the other. Those phrases were by no means habitually employed in 1896, either as informal appeals or in the capitalized versions of political sloganeering destined to become common in the twentieth century. But the values underlying the concepts authentically guided the campaigns of 1896 in ways that imparted enduring meaning to the outcome of their competition.

When he was not talking specifically about silver coinage, Bryan actually used the idea of "the people" as a centerpiece of many of his political speeches. When in Chicago, he said:

> As I look into the faces of these people and remember that our enemies call them a mob, and say they are a menace to free government, I ask: Who shall save the people from themselves? I am proud to have on my side in this campaign the support of those who call themselves the common people. If I had behind me the great trusts and combinations, I know that I would no sooner take my seat than they would demand that I use my power to rob the people in their behalf.

To many Americans the idea of "the people" represented the very foundation of democratic politics, and many thousands believed it had genuine meaning in the context of the Bryan campaign. But in 1896 the idea was even more specific, for it described not just "the people" in the abstract but a specific "people's movement" that had pressed itself upon the national consciousness, energized the silverites, and generated the preconditions for reform influence in the Democratic Party. Because it was not clear to the nation that the people's movement itself had been destroyed—its cooperatives crushed and its political party co-opted—Bryan came to symbolize its enduring life. This explained why Clarence Darrow, Eugene Debs, and many other Populists who had no illusions about the healing powers of the silver crusade ultimately came to join the "Great Commoner." They hoped he could rally the people to a new sense of their own prerogative and stimulate them, in L. L. Polk's old phrase, to "march to the ballot box and take possession of the government." To the extent that the silver crusade made much sense at all, it was in this symbolic context. The stirring rhythms of Bryan's "Cross of Gold" speech had energized the delegates at the Democratic convention—perhaps he could stir the American people as well. In the autumn of 1896, the hope was there, and this hope gave the Bryan campaign the deepest meaning it possessed.

Given the ballot box potentiality of "the people" as against "the great trusts and combinations," Republicans obviously could not afford to have the campaign decided on that basis. The countervailing idea of the "progressive society" materialized slowly out of the symbolic values embedded in the gold standard. The "sanctity of contracts" and "the national honor," it soon became apparent, were foremost among them. But, gradually, and with the vast distributional range afforded by the Republican campaign treasury, broader themes of "peace, progress, patriotism, and prosperity," came to characterize the campaign for William McKinley. The "progressive society" advanced by Mark Hanna in the name of the corporate community was inherently a well-dressed, churchgoing society. The various slogans employed were not mere expressions of a cynical politics, but rather the authentic assertions of an emerging American world view.

5

From a Populist perspective, the contest between "the people" and "the progressive society" was, in a practical sense, wholly irrelevant to the real purposes of the reform movement. Indeed, the narrow controversy over the "intrinsic value" of two competing metallic currencies was an affront to greenbackers. Not that they could do much about it; Populism had ceased to be an active force in American politics from the moment the third party had sacrificed its independent presence at the July convention.

The People's Party was therefore not a causative agent of anything significant that occurred during the frenzied campaign between the goldbugs and the silverites. From James Weaver to

Tom Watson, from Nebraska fusionists to Texas mid-roaders, the People's Party had become a reactive agent, responding as best it could to the initiatives of others. This being far from the proper business of an autonomous democratic movement struggling to break through a hierarchical party system, Populism quietly dissolved in the fall of 1896 as the morale of its two million followers collapsed. Democratic managers treated their Populist "allies" with such studied contempt that, for two months following the summer conventions, Marion Butler, the new Populist national chairman (replacing the discredited Taubeneck) was afraid to officially notify Bryan of his nomination for fear it would be publicly rejected. On the state level, fusion obliterated the party across the South where the standard-bearer of its bitter enemy, the Democratic Party, also graced the top of the Populist ticket. It is necessary to trace the actions of only one man—Tom Watson—during the autumn campaign of Populism to reveal the utter chaos that had descended upon the reform movement in every corner of the nation.

In his inaugural campaign speech in Atlanta early in August, Watson began actively pressuring the Democratic Party to remove Sewall from the ticket: "You cannot fight the national banks with any sincerity with a national banker as your leader." Butler reacted negatively to the speech, not because he wanted Sewall to remain but because he had been intimidated and outmaneuvered by the Democratic national chairman and by the silver lobby and, as a result, had lost a sense of his own prerogatives and those of the party he headed. Butler's course—to "do nothing of doubtful propriety while matters are happily shaping themselves in our favor"—betrayed a fatal lack of political self-confidence. Watson thereupon took a remarkable step that revealed the extent of his mistrust of fusion leaders. In interviews with the New York press, he attacked them publicly.

> If the National Convention at St Louis did not mean that Messrs. Bryan and Watson should be notified, why was a committee appointed to notify them? Why does Senator Allen, the chairman of the Committee, refuse to do what the convention instructed him to do? Is he afraid Mr. Bryan will

repudiate our support? If so, our party has a right to know that fact. If Mr. Bryan is ashamed of the votes which are necessary to elect him, we ought to know it.

By the standards of any self-respecting political party, these observations were eminently reasonable. They also were judged by the Republican press of the East to be highly newsworthy, as the intra-party quarrel among Populists had the effect of complicating Bryan's position as well.

Though Marion Butler, understandably, had cooled to the idea of a highly visible Watsonian presence in the fall campaign, the Georgian nevertheless embarked on a trip to Texas and Kansas early in September. The political atmosphere Watson found in Texas was hardly conducive to the kind of "patience and moderation" for which Butler had labored. A Watson intimate reported to Butler from Dallas that "Texas we find ripe for revolt. . . . The Pops are solid here against Bryan and Sewall." Indeed, a substantial number of the Texas third party leaders regarded the Bryan candidacy as a positive evil. Not only were they concerned that Populist support of Bryan would destroy the identity of the third party, but they also believed that the campaign's emphasis on free silver would undermine the basic monetary reforms of Populism. The Texans were so disturbed that they seriously considered the endorsement of McKinley as the "least of evils" in the 1896 campaign. When black Republican leaders in Texas offered to support the Populist state ticket in return for third party support of the Republican presidential ticket, the Texas mid-roaders took the matter under close advisement. Rumors of these discussions were in the air when Watson arrived in the state, and he responded warmly to the convivial Populist environment. "You must burn the bridges if you follow me," he asserted. To wild cheering, he announced his belief in "Straight Populism," for he did "not propose to be carried to one side of the road or the other." The campaign, he added, was "a movement of the masses. Let Bryan speak for the masses and let Watson speak for the masses and let Sewall talk for the banks and the railroads." Sewall was "a wart on the party. He is a knot on a log. He is a dead weight on the ticket." For good measure, Watson brought Butler and the national fusion leadership of the third party into his line of fire. The Texans, who shared most of those opinions and had a few others of their own, roared their approval.

While the reverberations from this speech rattled through the ranks of the silverites, Watson's itinerary took him to Kansas. There the Populist vice presidential candidate walked into Populist state headquarters under streaming banners that proclaimed, "Bryan and Sewall." As if to heighten the insult, Watson found that the very front of the Populist campaign office was decorated with huge portraits of the two Democrats whom the Kansas party took to be its national standard-bearers. The action in St. Louis by the politician-led Kansas party—one of full public support for Bryan and Sewall—was consummated back home through an arrangement with Democrats who exchanged Populist support of Sewall electors for Democratic support of the Populist state ticket. In speaking in Kansas, Watson was addressing party leaders who had sacrificed him in the name of their own local and state campaigns. The atmosphere was, to say the least, tense. As much as any man in America who responded to the call of reform launched by the Farmers Alliance, Watson had labored tirelessly in the people's movement. In Kansas in 1896 he gave his answer to the idea of fusion:

> Someone else must be asked to kill that Party; I will not. I sat by its cradle; I have fought its battles; I have supported its principles since organization . . . and don't ask me after all my service with the People's Party to kill it now. I am going to stand by it till it dies, and I want no man to say that I was the man who stabbed it to the heart. . . . No; Sewall has got to come down. He brings no votes to Bryan. He drives votes away from Bryan. . . . My friends, I took my political life in my hands when I extended the hand of fellowship to your Simpsons, your Peffers, your Davises in Georgia. The Georgia Democrats murdered me politically for that act. I stood by your men in Congress when others failed. I have some rights at the hands of Kansas. I have counted on your support. Can I get it?

The Populist audience responded with enthusiasm, pressed around the rostrum, and surrounded his carriage as he prepared to leave. Touched, Watson wrote his wife, "it is quite apparent that the rank and file of our party in Kansas are all right and will vote against their leaders if they get the chance."

6

The Kansas Populists who swarmed around Watson's carriage, like Watson himself, like the old Texas Alliancemen, had too many memories to acquiesce comfortably in the fusion politics of 1896. The spirit of Populism possessed meaning because of these memories. Indeed, collectively, they constituted the essence of the agrarian revolt: memories of farmers on the crop lien bringing food to striking railroad workers; of Alliancemen wearing suits of cotton bagging during the long war against the jute trust in the South; of the great lecturing campaigns to organize the South in 1887–89 and the nation in 1890–91; of mile-long Alliance wagon trains and sprigs of evergreen symbolizing the "living issues" in Kansas; of the formation of the "Alliance Aid" association to bring the Dakota plan of self-help insurance to the full agrarian movement; of Leonidas Polk, the Southern Unionist, announcing the end of American sectionalism to the cheers of Michigan farmers; of a day in Florida, "under the orange and banana trees," when the National Reform Press Association was created; of the time in South Carolina, so brief in that state, when impoverished farmers rebuked Ben Tillman to his face and tried to set a course different from his; of boys carrying torches in country parades for Tom Watson and girls knitting socks for "Sockless" Jerry Simpson; of the sprawling summer encampments that somehow seemed to give substance to the strange, inspiring, ethical vision of Tom Nugent; of John Rayner's lieutenants speaking quietly and earnestly in the houses of black tenants in the piney woods of East Texas; and, perhaps most symbolic of all, of men and women "who stood on chairs and marched back and forth cheering and crying" in response to the initial public reading of the "Second Declaration of American Independence" on a summer day in 1892 in the city of Omaha. It was this spirit—a collective hope for a better future, it seems—that animated American Populism, and it was these vibrant moments of shared effort that provided the evidence of its vitality, its aspirations, and its defeats.

And finding no coherent way to express itself in fusion, it was this spirit that expired in the autumn of 1896. Following his fiery speeches in Texas and his confrontations with the Kansans, Tom Watson sojourned briefly in the citadel of fusion in Nebraska and then, utterly demoralized, went home to Georgia for the remainder of the campaign. As the sole surviving symbol of a national Populist presence, he had become an anachronism. No one knew this more deeply than Watson or understood more fully what it meant for the future of the third party. The 1896 campaign had to do with the mobilization of new customs that were to live much more securely in American politics than the dreams of the Populists. That many of those customs were precisely the ones that so deeply disturbed the agrarian reformers constituted one of the more enduring ironies of the Populist experience.

7

The most visible difference in the efforts of the three parties in 1896 turned on money—not as a function of currency, but rather as the essential ingredient of modern electioneering. The Populist national treasurer wrote despairingly to Marion Butler that he was receiving less than a dozen letters a day containing "twenty-five cents to a dollar" from the demoralized Populist faithful throughout the nation. The Populist national campaign was literally almost penniless, and Butler found it necessary to establish his Washington headquarters in a building housing the political arm of the silver lobby, the National Silver Party. Only a belated, if humiliating, subsidy of $1000 from the Democratic national committee enabled the Populists to keep going until election day.

The Democratic campaign, although elaborately financed by Populist standards, was also run on a shoestring. The Republican press made great capital out of a supposed massive flow of funds from Western silver mineowners cascading into Democratic coffers. But the relatively modest sums that actually materialized went to the lobbying institutions previously created by the mineowners. Democrats worked desperately to place speakers in the field, but the shortage of money gradually channeled this effort toward the recruiting of self-supporting volunteers. The best educational force for the Democratic cause was Bryan himself. Through the early part of the campaign Bryan was forced to travel by commercial carrier, a circumstance that placed the entire presidential campaign at the mercy of local railroad timetables and earned the "boy orator" considerable ridicule in the goldbug press. The Democratic national committee was ultimately able to provide Bryan with a special train for the closing weeks of the campaign, but the silver crusade never quite lost the ad hoc character that had marked its inception.

In contrast, the massive national campaign for "honest money" engineered by Mark Hanna set a model for twentieth-century American politics. While the Democrats struggled to find volunteer speakers to tour the crucial states of the Midwest, the Republican campaign placed hundreds of paid speakers in the field. Individual contributions from wealthy partisans sometimes exceeded the entire amount the Democrats raised in their national subscription drive. Offerings from corporations, especially railroad corporations, reached even larger sums. Receipts and expenditures soared into the millions.

Mark Hanna presided over both the Chicago and New York campaign operations, coordinating an elaborate system of printing and distribution that involved many millions of pamphlets, broadsides, and booklets. So controlled and centralized was the Republican effort that the Chicago managers also took it upon themselves to assist in the supervision of state and local campaigns. To add a certain heft to their admonitions, the Chicago office dispatched almost a million dollars to various state organizations. The New York headquarters, focusing on the safe Eastern states, reported expenses of an additional $1,600,000.

The nation's new business combinations headquartered in New York largely financed the effort. Standard Oil contributed $250,000, a figure matched by J. P. Morgan. Hanna and railroad king James J. Hill were seen in a carriage "day after day," going from Wall Street to the office of the New York Central and the Pennsylvania railroads. Hanna repeatedly importuned the pres-

ident of New York Life, who just as frequently responded. The
corporate contributions mobilized in behalf of the 1896 Republican campaign for McKinley financed America's first concentrated mass advertising campaign aimed at organizing the minds
of the American people on the subject of political power, who
should have it, and why.

So supported logistically, the cultural politics of 1896 soon
unfolded in behalf of the "progressive society." Republican
references to the national honor extended to the party's role in
the Civil War. The "bloody shirt" waved in 1896; the fading
rhythms of Civil War loyalty were evoked with a measure of
subtlety, but evoked nevertheless. From his front porch in
Canton, William McKinley framed the larger issues in cultural
terms that looked to the past.

> Let us settle once for all that this government is one of honor
> and of law and that neither the seeds of repudiation nor
> lawlessness can find root in our soil or live beneath our flag.
> That represents all our aims, all our policies, all our purposes.
> It is the banner of every patriot, it is, thank God, today the
> flag of every section of our common country. No flag ever
> triumphed over it. It was never degraded or defeated and will
> not now be when more patriotic men are guarding it than
> ever before in our history.

In such a manner, the Republican Party first and foremost
moved to guard its basic constituency—one that had been created
by the war and had been solidified by repeated reminders of the
patriotism implicit in that initial allegiance. The politics of
sectionalism had always served this primary objective—to ensure
the party's organic constituency against anything that might hint
at wholesale apostasy. But in 1896, with the war receding in
time and with elections more and more depending on the votes
of people who had grown to maturity since Appomattox, the
appeal of the "bloody shirt" was boldly employed toward an
even more lasting objective—to merge the Republican Party's
past defense of the nation with contemporary emotions of
patriotism itself. Such an approach promised to elevate sectional
memories to national ones, forging a blend of the American
flag and the Grand Old Party that might conceivably cement a
bond of enduring vitality.

While William Jennings Bryan talked with passion and imprecision about the free coinage of silver, American flags—
literally millions of them—became the symbols of the struggle
to preserve the gold standard. McKinley himself became the
nation's "patriotic leader." The Republican campaign committee
purchased and distributed carloads of flags throughout the
country and Hanna conceived the idea of a public "flag day" in
the nation's leading cities—a day specifically in honor of William
McKinley. "Sound Money clubs" of New York and San Francisco
were put in charge of enormous flag day spectacles and supporting organizational work was carried forward with unprecedented attention to detail. When no less than 750,000 people
paraded in New York City, the *New-York Tribune* soberly reported—thirty-one years after Appomattox—that "many of
those who marched yesterday have known what it is to march
in war under the same flag that covered the city in its folds
yesterday all day long." In the critical Midwestern states, Civil
War veterans known as the "Patriotic Heroes" toured with
buglers and a cannon mounted on a flatcar. Slogans on the train
proclaimed that "1896 is as vitally important as 1861." So effective
was the Republican campaign that frustrated Democrats found
it difficult to show proper respect for the national emblem
without participating in some kind of public endorsement of
McKinley. Inevitably, some Democrats tore down Republican
banners—the American flag. Such actions did not hurt the
Republican cause.

McKinley adroitly yoked "free trade" to "free silver" as twin
fallacies threatening the orderly foundations of commerce as
well as the morality of the Republic. These threats were all
"Bryanisms" and collectively they added up to "anarchy." The
Republican antidotes thus consisted not merely of "sound
money" which protected both the sanctity of contracts and the
nation's honor, but equally symbolic appeals to the hope of
industrial workers for relief in the midst of a depression. To
drive this thought home, only one more slogan had to be added
to the litany marshaled by the Republicans. The phrase duly
materialized and was affixed to McKinley himself. He became
"the advance agent of prosperity." Hanna's New York money
and Dawes' Chicago printing presses ensured exposure of the

slogan throughout the nation. It often appeared emblazoned on huge banners flanked by phalanxes of American flags, the entire panoply carried high in the air by uniformed Civil War veterans wearing "sound money" buttons. Such broadly gauged cultural politics completely overwhelmed the vague call for free silver carried to the country by one barnstorming presidential candidate and a few platoons of volunteer Democratic speakers. In sheer depth, the advertising campaign organized by Mark Hanna in behalf of William McKinley was without parallel in American history. It set a creative standard for the twentieth century.

8

The election itself had an unusual continuity. Some students of the 1896 campaign have concluded that the enthusiasm for Bryan following his "Cross of Gold" speech was such that he would have swept to victory had the election been held in August. After that initial alarm the Republican organization quickly set in place the foundations for the mass campaign that followed, and by October the organizational apparatus assembled by Mark Hanna had clearly swung the balance to the Republicans. Yet the election results appeared fairly close, McKinley receiving 7,035,000 votes to Bryan's 6,467,000. But the Republicans had swept the North. While their margin in the Midwest was not overwhelming, it was a region Cleveland had carried for the Democrats four years before and one Bryan had been supremely confident of winning only a month before the election. Indeed, the results in the Midwest destroyed the party balance that had persisted since the Civil War, thus vastly changing national politics for the forseeable future. In fact, a cataclysm had befallen the Northern Democratic Party. Its progressive symbol in the Midwest, Governor John Peter Altgeld of Illinois, had suffered a surprising defeat, and state party tickets elsewhere in the North had been thrashed. The political appeal supposedly implicit in the idea of "the people" had received a powerful defeat.

It took awhile for the full implications to become apparent. Though it was not immediately noticed, the mature and victorious party of business had muted almost completely the egalitarian ideas that had fortified the party's early abolitionist impulses; the party of "peace, progress, patriotism and prosperity" had become not only anti-Irish, but anti-Catholic and anti-foreign generally. Its prior political abandonment of black Americans had quietly become internalized into a conscious white supremacy that manifested itself through a decreasing mention of the antislavery crusade as part of Republican services to the nation during the Civil War. The assertive party of business that consolidated itself in the process of repelling "Bryanism" in 1896 was, in a cultural sense, the most self-consciously exclusive party the nation had ever experienced. It was white, Protestant, and Yankee. It solicited the votes of all non-white, non-Protestant, or non-Yankee voters who willingly acquiesced in the new cultural norms that described gentility within the emerging progressive society. The word "patriotic" had come to suggest those things that white, Protestant Yankees possessed. This intensely nationalistic and racially exclusive self-definition took specific forms in 1896. The Democratic Party was repeatedly charged with being "too friendly" to foreigners, immigrants, and "anarchists." Indeed, the enduring implications echoed beyond the given tactics of a single campaign to define the restricted range of the progressive society itself. But for many of those who spoke for "the people," and for even greater numbers of the people themselves, no amount of fidelity to the new cultural values could provide entry to that society. While black Americans were to learn this truth most profoundly, its dimensions extended to many other kinds of "ethnic" Americans as well as to a number of economic groups and to women generally. The wall erected by the progressive society against "the people" signaled more than McKinley's victory over Bryan, more even than the sanctioning of massive corporate concentration; it marked out the permissible limits of the democratic culture itself. The "bloody shirt" could at last be laid away: the party of business had created in the larger society the cultural values that were to sustain it on its own terms in the twentieth century.

As an immediate outgrowth of the Bryan campaign, those

who extolled the doctrines of progress through business enterprise acquired a greater confidence, while those who labored in
behalf of "the people" suffered a profound cultural shock. A
number of the influential supporters of William McKinley, newly
secure in their prerogatives after the election, advanced from
confidence to arrogance; not only the nation, but the very forms
of its economic folkways had become theirs to define. In contrast,
a number of the followers of William Jennings Bryan, their idea
of America rebuked by the electorate, became deferential, either
consciously or unconsciously. The idea that serious structural
reform of the democratic process was "inevitable" no longer
seemed persuasive to reasonable reformers. Rather, it was
evident that political innovations had to be advanced cautiously,
if at all, and be directed toward lesser objectives that did not
directly challenge the basic prerogatives of those who ruled. The
thought became the inherited wisdom of the American reform
tradition, passed from one generation to another. A consensus
thus came to be silently ratified: reform politics need not concern
itself with structural alteration of the economic customs of the
society. This conclusion, of course, had the effect of removing
from mainstream reform politics the idea of people in an
industrial society gaining significant degrees of autonomy in the
structure of their own lives. The reform tradition of the twentieth
century unconsciously defined itself within the framework of
inherited power relationships. The range of political possibility
was decisively narrowed—not by repression, or exile, or guns,
but by the simple power of the reigning new culture itself.

In the election aftermath, William Jennings Bryan placed the
blame for his defeat on the weakness of the silver issue. In the
next four years he was to search desperately for a new issue
around which to rally "the people" against "the plutocracy." But
the decisive shift of voters to the Republican Party that had first
occurred in 1894 represented considerably more than a temporary reaction to the depression of 1893. Not only had the
new Republican majority been convincingly reaffirmed in 1896,
it was to prove one of the most enduring majorities in American
political history. Only the temporary split in the Republican
Party in 1912 was to flaw the national dominance that began in

1894 and persisted until the Great Depression of the 1930's. In the narrowed political world of the new century, the "Great Commoner" was never to locate his saving issue.

Meanwhile, the unraveling of the fabric of the People's Party between July and November, while fulfilling the predictions of mid-roaders, left little residue other than the bruised feelings and recriminations that Populists inflicted upon one another. Though the Kansas fusionists were rewarded for their efforts by achieving re-election on the joint Democratic-Populist ticket of 1896, it was only at the cost of the deterioration of the third party organization. Even before they could mount another campaign in 1898, Populist spokesmen in Kansas were conceding "the passing of the People's Party" and acknowledging that the agrarian crusade was over. That crusade, of course, had never really come to Nebraska. The stance of the Nebraska third party in 1896—essentially indistinguishable from that of the Nebraska Democratic Party—made the final passing of the vestiges of the Nebraska movement difficult to fix in time. The Nebraska fusion ticket of both 1897 and 1898 was, in any case, dominated by Democrats. In North Carolina, the election reforms passed by the Populist-Republican legislature of 1895 led to the election of number of black Republicans in 1896, setting the stage for a violent Democratic campaign of white supremacy in 1898. Almost total black disfranchisement resulted as the Democratic Party swept triumphantly back into power. In Texas, Populism polled almost a quarter of a million votes in November 1896— indicating that the third party was still growing at the moment its ideological and organizational roots were severed. The margin of Democratic victory was provided through the intimidation of Mexican-American voters and the terrorizing of black voters. Armed horsemen rode through the ranks of Negroes in Populist John B. Rayner's home county in East Texas and destroyed with force the years of organizing work of the black political evangelist. Elsewhere in the South, fusion obliterated the third party. As Watson put it, "Our party, as a party, does not exist any more. Fusion has well nigh killed it."

Pressed to its extremities, the Southern Democracy, "Bryanized" or not, revealed once again that one of its most enduring

tenets was white supremacy. It was the one unarguable reality
that had existed before, during, and after the Populist revolt.
Whatever their individual styles as dissenters from the Southern
Way, Mann Page in Virginia, Marion Butler in North Carolina,
Tom Watson in Georgia, Reuben Kolb in Alabama, Frank
Burkitt in Mississippi, Hardy Brian in Louisiana, W. Scott Morgan
in Arkansas, and "Stump" Ashby in Texas all learned this
lesson—as their "scalawag" predecessors had learned it before
them. But, as black Americans knew, white supremacy was a
national, not just a Southern, phenomenon. The progressive
society was to be a limited one.

9

North and South, Republican and Democratic, the triumphant
new politics of business had established similar patterns of public
conduct. Central to the new ethos was a profound sense of
prerogative, a certainty that in the progressive society only
certain kinds of people had a right to rule. Other kinds of
people, perforce, could be intimidated or manipulated or dis-
franchised. Students of the 1896 campaign have agreed on the
fact of overt employer intimidation of pro-Bryan factory workers
into casting ballots for McKinley. The pattern was especially
visible in the pivotal states of the Midwest. But the fact that the
same customs flourished on an even grander scale in the South,
where they were applied by Democrats to defraud Populists and
Republicans, pointed to some of the more ominous dimensions
of the emerging political exclusiveness. Both major parties were
capable of participating in the same political folkways of election
intimidation because they were both influenced by the same
sense of prerogative at the center of the emerging system of
corporate values.

For increasing numbers of Americans the triumph of the
business credo was matched, if not exceeded, by a conscious or
unconscious internalization of white supremacist presumptions.
Coupled with the new sense of prerogative encased in the idea
of progress, the new ethos meant that Republican businessmen
could intimidate Democratic employees in the North, Democratic

businessmen could intimidate Populists and Republicans in the South, businessmen everywhere could buy state legislators, and whites everywhere could intimidate blacks and Indians. The picture was not pretty, and it was one the nation did not reflect upon. In the shadow it cast over the idea of democracy itself, however, this mode of settling political disputes, or of making money, embedded within the soul of public life new patterns of contempt for alternative views and alternative ways of life. The cost of running for office, coupled with the available sources of campaign contributions and the increasing centralization of news gathering and news reporting, all pointed to the massive homogenization of business politics.

In addition to the banishment of the "financial question" as a political issue, three other developments soon materialized in the wake of the 1896 election to establish enduring patterns for the twentieth century—the rapid acceleration of the merger movement in American industry, the decline of public participation in the democratic process itself, and corporate domination of mass communications. Corporate America underwent periodic waves of heightened consolidation, from 1897 to 1903, 1926 to 1931, and 1945 to 1947. Building upon prior levels of achievement, the process accelerated once again in the 1960's. The "trusts and combinations" that the Populists believed were "inherently despotic" rode out the brief popular clamor for antitrust legislation before World War I. Such largely marginal reforms as were able to run the lobbying gantlet within the United States Congress were vitiated by subsequent Supreme Court decisions. To the despair of antitrust lawyers in the Department of Justice, the combination movement became a historical constant of twentieth-century American life as a structure of oligopoly was fashioned in every major industry. At the same time, "bank holding companies" had fashioned networks of corporate consolidation scarcely imagined by Populist "calamity howlers." Within the narrowed range of political options available to twentieth-century advocates of reform, however, this seminal development within the structure of finance capitalism was not a matter of sustained public debate.

The passing of the People's Party left the Southern Democracy

securely in the hands of conservative traditionalists in every state
of the region. There, "election reform" proved to have as many
dimensions as "banking reform" did nationally. The process
began with the disfranchisement of blacks in Mississippi in 1890
and accelerated after 1896 as state after state across the South
legalized disfranchisement for blacks and made voting more
difficult for poorer whites. The movement for election reform
was accompanied by a marked decline in the relationship of
public issues to the economic realities of Southern agriculture,
a persistent twentieth-century folkway that contributed to a
sharp drop in popular interest in politics among Southerners. In
the twentieth century, the voting percentages in the South
remained far below the peaks of the Populist decade. Within
the narrower permissible limits of public disputation, more and
more Americans felt increasingly distant from their government
and concluded that there was little they could do to affect
"politics." The sense of personal participation that Evan Jones
had felt in 1888 when he summoned 250,000 Alliancemen to
meet in more than 175 separate courthouses—without notifying
anyone in the courthouses—pointed to a kind of intimacy
between ordinary citizens and their government that became
less evident in twentieth-century America. To Alliancemen, it
was *their* courthouse and *their* government.

The demise of the National Reform Press Association, meanwhile, ended a journalistic era that dated back to colonial times,
an era that might be characterized as seat-of-the-pants democratic journalism. Volatile, spirited, always opinionated, the one-
man journals of eighteenth- and nineteenth-century America
also provided the nation with a counterforce to any and all
orthodoxies. But as centralization came to all phases of American
industry, it also came to newspapers. Gradually, inexorably,
journalism became a corporate business, one in which competition in ideas inevitably diminished. With mergers and "chains,"
the newspaper business gradually became, like other aspects of
commerce, big business.

The political implications were large. By the last quarter of
the twentieth century so intimidated were the nation's print and
electronic journalists by the world of corporate power they lived

in and worked for that corporate domination of the United States Congress was not a political fact that "the media" ever succeeded in making clear to the American public. Though corporate domination of the legislative process had been the governing political folkway of every decade of the twentieth century, the fact itself seemed beyond the capacity of modern corporate journalists to report with clarity. Simply stated, the cultural prerequisites for them to do so had not been fulfilled. Granted that the psychological needs of participating in a "free press" were such that owners, editors, and reporters alike occasionally worked zealously at uncovering various influence scandals, the continuing structure of corporate domination was not something that formed the centerpiece of political journalism in the United States.*

While it required a daily exposure only to one six-month session of any American legislature, state or national, to discover the pervasiveness of corporate influence, few Americans were in a position to take six months off from their jobs to acquire the necessary political education. The popular innocence concerning the real structure of parliamentary democracy that resulted from these unperceived power relationships became, in fact, the strongest stabilizing factor supporting the political status quo in the corporate state. With such innocence routinely governing popular perceptions of politics and democracy in America,

* I do not intend here to question the ethics of the nation's journalists; indeed, the point is that their role in this process is a minor one. The substance of American political reporting turns less on the ethics of individual reporters than on the structure of cultural restraints under which they labor. The simple historical fact is that the political values that reporters might impugn are the reigning values of the newspaper and television corporations they work for. These cultural imperatives, while awkward, are unarguably germane to any assessment of post-Populist journalism in America. Given the cultural barriers to journalistic clarity, the individual tenacity required to perform a professional investigative job on congressional lobbying would necessarily create the kind of furor among advertisers that is not conducive to successful newspaper or television careers. The issue is one of power, not ethics. As presently constituted, "news reporting" in the era of the corporate state is a corporate business— inherently hierarchical in form, inherently undemocratic in product. Few Americans know these causal relationships more intimately than the reporters themselves.

"public opinion" became something less than an effective democratic safeguard.

Collectively, these patterns of public life, buttressed by the supporting faith in the inevitability of economic progress, ensured that substantive democratic political ideas in twentieth-century America would have great difficulty in gaining access to the progressive society in a way that approached even the marginal legitimacy achieved by Populist "calamity howlers." The American populace was induced to accept as its enduring leadership a corporate elite whose influence was to permeate every state legislature in the land, and the national Congress as well. A new style of democratic politics had become institutionalized, and its cultural boundaries were so adequately fortified that the new forms gradually described the Democratic Party of opposition as well as the Republican Party of power. A critical cultural battle had been lost by those who cherished the democratic ethos. The departure of a culturally sanctioned tradition of serious democratic reform thought created self-negating options: reformers could ignore the need for cultural credentials, insist on serious analysis, and accept their political irrelevance as "socialists"; or they could forsake the pursuit of serious structural reform, and acquire mainstream credentials as "progressives" and "liberals." In either case, they could not hope to achieve what Populists had dared to pursue—cultural acceptance of a democratic politics open to serious structural evolution of the society. The demonstrated effectiveness of the new political methods of mass advertising meant, in effect, that the cultural values of the corporate state were politically unassailable in twentieth-century America.

10

The socialists who followed the Populists did not really understand this new reality of American culture any more than most Populists did in the 1890's. To whatever extent socialists might speak to the real needs of "the people"—as they intermittently did for two generations—and however they might analyze the destructive impact of the progressive society upon the social relations and self-respect of the citizenry—as they often did during the same period—the advocates of popular democracy who spoke out of the socialist faith were never able to grapple successfully with the theoretical problem at the heart of their own creed. While the progressive society was demonstrably authoritarian beyond those ways that Thomas Jefferson had originally feared, and while it sheltered a party system that was intellectually in homage to the hierarchical values of the corporate state itself—cultural insights that provided an authentic connecting link between Populism and socialism—the political power centered in "concentrated capital" could not be effectively brought under democratic control in the absence of some correspondingly effective source of non-corporate power. While the Populists committed themselves to a people's movement of "the industrial millions" as the instrument of reform, the history of successful socialist accessions to power in the twentieth century has had a common thread—victory through a red army directed by a central political committee. No socialist citizenry has been able to bring the post-revolutionary army or central party apparatus under democratic control, any more than any non-socialist popular movement has been able to make the corporate state responsive to the mass aspirations for human dignity that mock the pretentions of modern culture. Rather, our numerous progressive societies have created, or are busily creating, over-powering cultural orthodoxies through which the citizenry is persuaded to accept the system as "democratic"—even as the private lives of millions become more deferential, anxiety-ridden, and (no other phrase will serve) less free.

Increasingly, the modern condition of "the people" is illustrated by their general acquiescence in their own political inability to affect their governments in substantive ways. Collective political resignation is a constant of public life in the technological societies of the twentieth century. The folkway knows no national boundary, though it does, of course, vary in intensity in significant ways from nation to nation. In the absence of alternatives, millions have concentrated on trying to find private modes of escape, often through material acquisition. Indeed, the operative standard of progress in both ideological worlds of socialism and capitalism focuses increasingly upon economic indexes. Older aspirations—dreams of achieving a civic culture grounded in generous social relations and in a celebration of the vitality of human cooperation and the diversity of human aspiration itself—have come to seem so out of place in the twentieth-century societies of progress that the mere recitation of such longings, however authentic they have always been, now constitutes a social embarrassment.

But while the doctrines of socialism have not solved the problem of hierarchical power in advanced industrial society, its American adherents can scarcely be blamed for having failed to build a mass popular following in the United States. Though they never quite achieved either a mass movement or a movement culture that matched the size, richness, and creativeness of the agrarian cooperative crusade, they were, in the aftermath of the cultural consolidation that accompanied the Populist defeat, far more politically isolated than their agrarian predecessors had been. To an extent that was not true of many other societies, the cultural high ground in America had been successfully consolidated by the corporate creed a decade before American socialists, led by Eugene Debs, began their abortive effort to create a mass popular base. The triumphant new American orthodoxy of the Gilded Age, sheltering the two-party system in a dialogue substantially unrelated to democratic structural reform of the inherited economic and social system, consigned the advocates of such ideas to permanent marginality. The Populists have thus been, to date, the last American reformers with authentic cultural credentials to solicit mass support for the idea of achieving the democratic organization of an industrialized society.

But while American socialists, for reasons they themselves did not cause, can be seen in retrospect as never having had a chance, they can be severely faulted for the dull dogmatism and political adolescence of their response to this circumstance. Though their primary recruiting problem turned on their lack of domestic cultural credentials—the working poor wanted justice, but they wanted it as loyal Americans—socialists reacted to continued cultural isolation by celebrating the purity of their "radicalism." Thus, individual righteousness and endless sectarian warfare over ideology came to characterize the politics of a creed rigidified in the prose of nineteenth-century prophets. As a body of political ideas, socialism in America—as in so many other countries—never developed a capacity for self-generating creativity. It remained in intellectual servitude to sundry "correct" interpretations by sundry theorists—mostly dead theorists—even as the unfolding history of the twentieth century raised compelling new questions about the most difficult political problem facing mankind: the centralization of power in highly technological societies. If it requires an army responsive to a central political committee to domesticate the corporate state, socialism has overwhelmingly failed to deal with the question of who, in the name of democratic values, would domesticate the party and the army. In the face of such a central impasse, it requires a rather grand failure of imagination to sustain the traditional socialist faith.

11

As a political culture, Populism fared somewhat better during its brief moment. Third party advocates understood politics as a cultural struggle to describe the nature of man and to create humane models for his social relations. In the context of the American ethos, Populists therefore instinctively and habitually resisted all opposition attempts, through the demagogic expedient of labeling new ideas as "radical," to deflect those ideas from serious discussion. As Populists countered this ploy, they defended the third party platform as "manly and conservative." This Populist custom extended especially to the movement's hard core of lecturers and theorists, who understood the Omaha Platform as a series of threshold demands, to be promptly augmented upon enactment, as the successful popular movement advanced to implement its ultimate goal of a "new day for the industrial millions." In articulating their own social theory, their cause of "education" could advance in step with each stage of democratic implementation, as the "plain people" gained more self-respect from the supportive culture of their own movement and as they gained confidence in their rights as citizens of a demonstratively functioning democracy. They never found the means to bring the educational power of their movement culture to remotely enough American voters, but if one thing may be said of the Populists, it is that they tried.

Populism in America was not the sub-treasury plan, not the greenback heritage, not the Omaha Platform. It was not, at bottom, even the People's Party. The meaning of the agrarian revolt was its cultural assertion as a people's movement of mass democratic aspiration. Its animating essence pulsed at every level of the ambitious structure of cooperation: in the earnest probings of people bent on discovering a way to free themselves from the killing grip of the credit system ("The suballiance is a schoolroom"); in the joint-notes of the landed, given in the name of themselves and the landless ("The brotherhood stands united"); in the pride of discovery of their own legitimacy ("The merchants are listening when the County Trade Committee talks"); and in the massive and emotional effort to save the cooperative dream itself ("The Southern Exchange Shall Stand"). The democratic core of Populism was visible in the suballiance resolutions of inquiry into the patterns of economic exploitation ("find out and apply the remedy"); in the mile-long Alliance wagon trains ("The Fourth of July is Alliance Day"); in the sprawling summer encampments ("A pentecost of politics"); and, perhaps most tellingly, in the latent generosity unlocked by the culture of the movement itself, revealed in the capacity of those who had little, to empathize with those who had less ("We extend to the Knights of Labor our hearty sympathy in their manly struggle against monopolistic oppression," and "The Negro people are part of the people and must be treated as such").

While each of these moments occurred in the 1890's, and have practical and symbolic meaning because they did occur, Populism in America was not an egalitarian achievement. Rather, it was an egalitarian attempt, a beginning. If it stimulated human generosity, it did not, before the movement itself was destroyed, create a settled culture of generosity. Though Populists attempted to break out of the received heritage of white supremacy, they necessarily, as white Americans, did so within the very ethos of white supremacy. At both a psychological and political level, some Populists were more successful than others in coping with the pervasive impact of the inherited caste system. Many were not successful at all. This reality extended to a number of pivotal social and political questions beside race—sectional and party loyalties, the intricacies of power relationships embedded in the monetary system, and the ways of achieving a politics supportive of popular democracy itself. In their struggle, Populists learned a great truth: cultures are hard to change. Their attempt to do so, however, provides a measure of the seriousness of their movement.

Populism thus cannot be seen as a moment of triumph, but as a moment of democratic promise. It was a spirit of egalitarian hope, expressed in the actions of two million beings—not in the prose of a platform, however creative, and not, ultimately, even in the third party, but in a self-generated culture of collective dignity and individual longing. As a movement of people, it was expansive, passionate, flawed, creative—above all, enhancing in its assertion of human striving. That was Populism in the nineteenth century.

But the agrarian revolt was more than a nineteenth-century experience. It was a demonstration of how people of a society containing a number of democratic forms could labor in pursuit of freedom, of how people could generate their own democratic culture in order to challenge the received hierarchical culture. The agrarian revolt demonstrated how intimidated people could create for themselves the psychological space to dare to aspire grandly—and to dare to be autonomous in the presence of powerful new institutions of economic concentration and cultural regimentation. The Omaha Platform gave political and symbolic substance to the people's movement, but it was the idea animating the movement itself that represents the Populist achievement. That idea—at the very heart of the movement culture—was a profoundly simple one: the Populists believed they could work together to be free individually. In their institutions of self-help, Populists developed and acted upon a crucial democratic insight: to be encouraged to surmount rigid cultural inheritances and to act with autonomy and self-confidence, individual people need the psychological support of other people. The people need to "see themselves" experimenting in new democratic forms.

In their struggle to build their cooperative commonwealth, in their "joint notes of the brotherhood," in their mass encampments, their rallies, their wagon trains, their meals for thousands, the people of Populism saw themselves. In their earnest suballiance meetings—those "unsteepled places of worship"—they saw themselves. From these places of their own came "the spirit that permeates this great reform movement." In the world they created, they fulfilled the democratic promise in the only way it can be fulfilled—by people acting in democratic ways in their daily lives. Temporary victory or defeat was never the central element, but simple human striving always was, as three epic moments of Populism vividly demonstrated in the summers of 1888, 1889, 1890. These moments were, respectively, the day to save the exchange in Texas, Alliance Day in Atlanta, when 20,000 farmers massed against the jute trust, and Alliance Day in Winfield, Kansas. Though L. L. Polk made a stirring speech to the Kansans on July 4, 1890 in Winfield, what he said was far less important than what his listeners were seeing. The wagon trains of farm families entering Cowley County from one direction alone stretched for miles. In this manner, the farmers saw their own movement: the Alliance was the people, and the people were together. As a result, they dared to listen to themselves individually, and to each other, rather than passively follow the teachings of the received hierarchical culture. Their own movement was their guide. Fragile as it was, it nevertheless opened up possibilities of an autonomous democratic life. Because this happened, the substance of American Populism went beyond the political creed embedded in the People's Party, beyond the evocative images of Alliance lecturers and reform editors, beyond even the idea of freedom itself. The Populist essence was less abstract: it was an assertion of how people can *act* in the name of the idea of freedom. At root, American Populism was a demonstration of what authentic political life can be in a functioning democracy. The "brotherhood of the Alliance" attempted to address the question of how to live. That is the Populist legacy to the twentieth century.

12

In their own time, the practical shortcoming of the Populist political effort was one the agrarian reformers did not fully comprehend: their attempt to construct a national farmer-labor coalition came before the fledgling American labor movement was internally prepared for mass insurgent politics. Alliance lecturers did not know how to reach the laboring masses in the nation's cities, and, in the 1890's, the labor movement could not effectively reach them either. Though a capacity for germane economic analysis and a growing sense of self developed in those years among American workers, their advances had not, by the time Populism arrived, been translated on a mass scale into practical political consciousness. In the 1890's, growing numbers of American workers were desperately, sometimes angrily, seeking a way out of their degradation, but the great majority of them carried their emotions with them as they voted for the major parties—or did not vote at all. By the time American industrial workers finally found a successful organizing tactic—the sit-down strike—in the 1930's, a sizable proportion of America's agricultural poor had been levered off the land and millions more had descended into numbing helplessness after generations of tenantry. Thus, when the labor movement was ready, or partly ready, the mass of farmers no longer were. That fact constitutes perhaps the single greatest irony punctuating the history of the American working class.

As for the farmers, their historical moment came in the late 1880's. They built their cooperatives, sang their songs, marched, and dreamed of a day of dignity for the "plain people."

But their movement was defeated, and the moment passed. Following the collapse of the People's Party, farm tenantry increased steadily and consistently, decade after decade, from 25 per cent in 1880, to 28 per cent in 1890, to 36 per cent in 1900, and to 38 per cent in 1910. The 180 counties in the South where at least half the farms had been tenant-operated in 1880 increased to 890 by 1935. Tenantry also spread over the fertile parts of the corn belt as an increasing amount of Midwestern farmland came to be held by mortgage companies. Some 49 per cent of Iowa farms were tenant-operated in 1935 and the land so organized amounted to 60 per cent of the farm acreage in the state. In 1940, 48 per cent of Kansas farms were tenant-operated. The comparable figure for all Southern farms was 46 per cent. But in the South, those who had avoided tenantry were scarcely in better condition than the sharecroppers. An authoritative report written by a distinguished Southern sociologist in the 1930's included the information that over half of all land-owners had "short-term debts to meet current expenses on the crop." The total for both tenants and landowners shackled to the furnishing merchant reached 70 per cent of all farmers in the South. As one historian put it, the crop lien had "blanketed" the entire region. As in the Gilded Age, the system operated in a way that kept millions living literally on the wages of peonage.

13

If the farmers of the Alliance suffered severely, what of the agrarian crusade itself? What of the National Farmers Alliance and Industrial Union, that earnest aggregation of men and women who had striven for a "cooperative commonwealth"? And what of the People's Party? Why, precisely, did it all happen, and what is its historical meaning?

As agrarian spokesmen were forever endeavoring to make clear to Americans—indeed, in S. O. Daws's early "history," published in 1887—the cooperative movement taught the farmers "who and where the enemies of their interests were." The Alliancemen who learned that lesson first were the men who had been sent out by the cooperatives to make contact with the surrounding commercial world, men like Daws, Lamb and Macune in Texas, and Loucks and Wardall in Dakota. Though they possessed different political views and different sectional memories, they were altered in much the same way by the searing experience of participating in, and leading, a thwarted hope. They became desperate, defensively aggressive, angry, and creative. They reacted with boycotts, with plans for mutual self-help insurance societies, with the world's first large-scale cooperatives, and with the sub-treasury plan. The marketing and purchasing agents who learned the lessons of cooperation became both movement politicians and ideological men.

They built their cooperatives, developed new political ideas, and fashioned a democratic agenda for the nation. The destruction of the cooperatives by the American banking system was a decisive blow, for it weakened the interior structure of democracy that was the heart of the cooperative movement itself. Though, in one final burst of creativity, the agrarian radicals were able to fashion their third party, that moment in Omaha in the summer of 1892 was the movement's high tide. There was no way a political institution—a mere party—could sustain the day-to-day democratic ethos at the heart of the Alliance cooperative.

And here, among these threads woven through the tapestry of the agrarian revolt, reposes the central historical meaning of Populism. It would, perhaps, constitute a fitting epitaph for the earnest farmers of the Alliance to place these threads in relation to one another and review them a final time. For while they say much about the meaning of the agrarian revolt, they also reveal a great deal about the world Americans live in today and—most important of all—about how modern people have been culturally organized to think about that world. The very appearance of such a pale version of Populism as the shadow movement dramatizes the fact that the Alliance organizers did not, could not, with a wave of the hand overcome all of the American cultural barriers to reform merely by successfully placing their new People's Party upon the stage of national politics in 1892. Indeed, an inspection of what the organizers of Populism did and did not achieve illustrates the elemental difficulties facing serious democratic reformers in any industrial society at any time.

It is clear, for example, that the shadow movement did not emerge merely because of the maneuverings of two amiable Nebraska politicians named Allen and Bryan, nor even because of the demonstrated capacity of silver money to infiltrate an impoverished reform movement. As with all democratic movements in industrial societies, the Populist cause faced hazards that were much more organic than could be revealed by the tactical twists and turns of aggressive office-seekers or aggressive silver magnates.

If the central task of democratic reform involves finding a way to oppose the received hierarchical culture with a newly created democratic culture, and if, as the Alliance experience reveals, progress toward this culminating climax necessarily must build upon prior stages of political and organizational evolution that have the effect of altering the political perspectives of millions of people, then democratic movements, to be successful, clearly require a high order of sequential achievement. Towering over all other tasks is the need to find a way to overcome deeply ingrained patterns of deference permeating the entire social order. For this to happen, individual self-respect obviously must take life on a mass scale. At the onset of this process for the Populists, small battles (the bulking of cotton) and larger battles (the war with the jute trust) needed to be fought and won so that the farmers attained the beginnings, at least, of collective self-confidence. Arrayed against these democratic dynamics was the continuing cultural authority embedded in received habits of thought, and the readily available ways that the press, the public school, and the church could refortify these inherited patterns. Against such powerful counter-influences, the only defense available to a democratic movement such as the Alliance lay within its own organizational institutions. Interior lines of communication were essential to maintain the embryonic and necessarily fragile new culture of mass self-respect engendered by the reform movement itself. In the absence of such continuing and self-generated democratic cultural influences, the organizers of popular movements inevitably face loss of control over their own destiny. They will be overwhelmed, not by their own party's politicians *per se*, or by passing corporate lobbies, but, more centrally, by the inherited culture itself. Their dreams will vanish into the maw of memory, as their impoverished constituencies, battered on all sides by cultural inducements to conform to received habits, gradually do precisely that. As the Alliance organizers understood, mass democratic movements, to endure, require mass democratic organization. Only in this way can individual people find the means to encourage one another through their own channels of communication—channels that are free from the specialized influences of the hierarchical society they seek to reshape.

In the years in which the National Farmers Alliance and

Industrial Union created, through its cooperative crusade, the movement culture of Populism, this interior channel of communication was centered in the Alliance lecturing system. It was this instrument of self-organization that permitted the hopes of masses of farmers to be carried forward to their spokesmen and allowed the response of the same spokesmen to go back to the "industrial millions." Democratic pressure from below—from crop-mortgaged farmers desperate to escape their furnishing merchants—emboldened the early Alliance leadership to undertake the "joint-note" plan within the centralized statewide cooperative. And tactical democratic strategy from above, in the form of William Lamb's "politics of the sub-treasury," helped raise the political consciousness of many hundreds of thousands of farmers to a level necessary for them to make the personal decision to break with their cultural inheritance and support the new People's Party. The Alliance lecturing system was organic to this democratic message-carrying, to and fro.

As the farmers labored to create a workable infrastructure of mass cooperation in 1887–92, the opposition of the American banking and corporate communities gradually brought home to Alliance leaders, and to masses of farmers, the futility of the cooperative effort—in the absence of fundamental restructuring of the monetary system. The Allianceman who was forced by events to learn these lessons first—Charles Macune—was the first to brood about this dilemma, and the first to formulate a democratic solution, the sub-treasury land and loan system.

The status that the sub-treasury plan came to have in reform ranks is revealing. For, to put the matter as quietly as possible, Macune's plan was democratic. Or, to put it in archaic political terminology, it was breathtakingly radical. Under the sub-treasury, the power of private moneylenders to decide who "qualified" for crop loans and who did not would have been ended. Similarly, the enormous influence of moneylenders over interest rates would also have been circumscribed. The contracted currency, the twenty-five year decline in volume and prices, would have been ended in one abrupt—and democratic—restructuring. The prosperity levels of 1865 would have been reclaimed in one inflationary—and democratic—swoop. Most important of all, the sub-treasury addressed a problem that has largely defeated twentieth-century reformers, namely the maldistribution of income within American society. By removing some of the more exploitive features embedded in the inherited monetary system, the sub-treasury would have achieved substantive redistribution of income from creditors to debtors. Put simply, a more democratic monetary system would have produced a more democratic sharing of the nation's total economic production. The "producing classes," no longer quite so systematically deprived of the fruits of their efforts, would have gotten a bit more of the fruits. Hierarchical forms of power and privilege in America would have undergone a significant measure of rearrangement.

As subsequent history was to reveal, these Populistic premises proved to be beyond the conceptual reach of twentieth-century Americans. Restructuring of American banking was not something about which New Dealers or New Frontiersmen could think with sustained attention. The received culture has proved to be so powerful that substantive ideas about a democratic system of money and credit have become culturally inadmissible. Such ideas (the sub-treasury concept of a treasury-based democratic bank will do adequately as an example) are, in the judgment of prevailing cultural authority, "unsound." No one disputes such culturally sanctioned wisdom today, any more than the goldbug simplicities were disputed in "informed" circles during the Gilded Age.

The shrinking parameters of twentieth-century reform thought thus help to underscore the Populist achievement. How was it that so many people in the 1890's came to associate themselves with such an inadmissible idea as the sub-treasury plan? We may be secure in the knowledge that the Populists were not "smarter" than modern Americans. Nor did it happen just because "times were hard." It happened because Populists had constructed within their own movement a specific kind of democratic environment that is not normally present in America. In the face of all the counterattacks employed by the nation's metropolitan press, culturally in step as it was with the needs of "sound-money" bankers, the sub-treasury plan was brought home to millions of Americans as a function of the movement culture of Populism. This culture was comprised of many ingredients—most visible being the infrastructure of local, county, state, and national Alliance organizations. But this was surface. More real were all of the shared experiences within the Alliance—the elaborate encampments, the wagon trains, the meals for thousands—and more real still were the years of laboring together in the suballiances to form trade committees, to negotiate with merchants, to build the cooperatives to new heights, to discuss the causes of adversity, and, in time, to come to the new movement folkway, the "Alliance Demand." The Demands took on intense practical meaning, first at Cleburne in 1886 and later as the St. Louis Platform of 1889, the Ocala Demands of 1890, and the Omaha Platform of 1892. Because these multiple methods of interior communication existed, Alliancemen found a way to believe in their own movement, rather than to respond to what the larger society said about their movement. In sum, they built insulation for themselves against the received hierarchical culture. Because they did so, the farmers of the Alliance overcame, for a time, their deference; they gained, for a time, a new plateau of self-respect that permitted autonomous democratic politics. Because of this evolution, which was essentially the evolution of Populism, they could dare to have significant democratic aspirations in ways that twentieth-century Americans have rarely been able to emulate—and never on such a scale as Populism. Because the sub-treasury plan promised (effectively, we now understand) to save the Alliance cooperative from banker strangulation, the new monetary system undergirded the cooperative idea: the Omaha Platform of the People's Party preserved the basic dreams of the Farmers Alliance.

But while these conclusions are self-evident in a political sense, and explain how the People's Party happened, they conceal an underlying organizational flaw that eventually undermined the reform movement. For while the cooperative idea awaited the enactment of a new and more democratic system of industrial commerce—the passage by a Populist Congress of the sub-treasury system—the basic cooperative structure of the Alliance

gradually disintegrated. The retrenchment of the Texas Exchange, from a centralized statewide marketing and purchasing
credit cooperative to a much smaller cash-only purchasing
cooperative, excluded the bulk of cashless Alliance farmers from
participation.* The Texas setback of 1889–90 was followed by
similar retrenchments at local and statewide levels throughout
the South and West in 1890–93. The underlying cause was
everywhere the same: lack of access to credit. But in some places,
where especially favorable circumstances temporarily bridged
the credit problem, cooperatives were destroyed by raw appli-
cations of commercial power. The highly successful multi-state
livestock marketing cooperative, conceived by Kansans and
exported to Missouri, Nebraska, and the Dakotas, was killed by
the simple decision of the Livestock Commission in Chicago to
refuse to deal with the farmer cooperative. The decision was
justified on the ground that the cooperative, in distributing
profits to its membership, violated the "anti-rebate" rule of the
commission! That was the end of that. But whether the cause
of cooperative defeat was the tactical influence of the American
financial community, control of access to credit, or simple power,
as in the case of the Chicago Livestock Commission, the orga-

---

* That fact illustrates in still another way the extent to which traditional class
analysis helps to obscure this central dynamic of the Populist movement. It
made little difference whether the number of tenants in a given state comprised
15 or 30 per cent of all farmers, or whether all, or half, of the tenants joined
the Alliance. The relevant "statistic" was the fact that an overwhelming majority
of all farmers, landowners and landless alike, were locked into the crop lien, the
discriminatory freight rate and the chattel mortgage.

Rochdale-type cash cooperatives could not help such farmers, whether they
still owned their land, were losing their land, had already lost their land, or had
never possessed any. If capitalist historians have missed the Populist movement
because of reliance on fragile economic determinism (the agrarian revolt resulted
from "hard times"), Marxist historians have similarly deluded themselves by
easy reliance on class categories (Populism could not be "progressive" because
the movement was dominated by "landowners"). One may perhaps be forgiven
for suggesting that the manifest lack of creativity in twentieth-century political
thought may be at least partly traceable to the remarkable complacency with
which categories of political description, sanctioned in the intellectual traditions
of socialism and capitalism, are employed as interpretive devices to "explain"
historical causation. If these categories do not work to reveal the largest mass
movement in nineteenth-century America, it is possible to wonder what else in
American history they do not reveal.

nizational structure of the Alliance became fatally weakened and
the mass cohesion of the movement was, consequently, fatally
compromised.

All important sectors of commercial America opposed the
cooperative movement, not only banks and commission agencies,
but grain elevator companies, railroads, mortgage companies,
and, perhaps needless to add, furnishing merchants. The National Farmers Alliance itself persisted as an institution, but the
cooperative purpose that sustained the personal day-to-day
dedication of mèmbers to their own institution did not persist.
Once the politics of the sub-treasury had been orchestrated
through the lecturing system in order to bring on the new party,
the lecturing system itself withered. The reason was a basic one;
the lecturers no longer had anything substantive to lecture
about. The Alliance could no longer save the farmers; only the
new party could bring the needed structural changes in the
American economic system.

"Lecturing" thus became a function of the People's Party. The
new lecturers who provided the continuing internal communications link within the movement culture were the reform
editors. The National Reform Press Association was to the
People's Party what the lecturing system was to the Alliance, the
interior adhesive of the democratic movement. The flaw in all
this was the simple fact that the National Reform Press Association did not have an organized constituency, as Alliance lecturers had earlier possessed. Within the People's Party, as it
organized itself, there could be no continuing democratic dialogue, no give and take of question and answer, of perceived
problem and attempted solution, between rank-and-file members and elected spokesmen, such as had given genuine democratic meaning to the days of cooperative effort within the
Alliance. Rather, reform editors asserted and defended the
Populist vision, and their subscribers, in organizational isolation,
received these views in a passive state, as it were. Such a dynamic
undermines the very prospect of sustaining a democratic culture
grounded—as it must be to be democratic—in individual self-
respect and mass self-confidence. Individual self-respect requires
self-assertion, the performance of acts, as farmers performed in

their suballiance business meetings, in their country trade committees, in their statewide marketing and purchasing co-operatives. But, in vivid contrast, the passive reading of reform newspapers fortified inherited patterns of deference. Thus the democratic intensity of the People's Party declined over time, because means were not found to sustain democratic input from the mass of participating Populists. In this fundamental political sense, the movement culture of Populism was not, and could not be, as intense as the movement culture originally generated within the Alliance cooperative crusade. The people were the same, but a crucial democratic element was no longer present.

The original Alliance organizers sensed this. They labored long and hard to keep the Alliance organization intact, even after they had successfully formed the People's Party. But within the intellectual traditions of reform theory in western culture, Alliancemen found few nineteenth-century models to guide them in this organizational endeavor. Indeed, in a number of quarters that were increasingly influential in vanguard political circles in Europe, the idea of mass democratic institutions serving as the central agent of social change was being rejected at the very moment the agrarian revolt appeared in America. To many would-be revolutionaries—even as they asserted their beliefs in the name of "democracy"—the rural peasantry and the urban proletariat had both demonstrated their inability to participate meaningfully in serious restructuring of industrial societies. The answer reached by Lenin was, of course, to entrust the advent of the "new day" to a small intellectual elite, tightly organized as a revolutionary party. In such a structure the risks of institutionalizing authoritarian rule are obvious today, if they were not then. At any event, such theoretical approaches clearly had no appeal to Alliance radicals, driven as they were by visions of authentic mass democracy. The Alliance founders, then, turned for their model not to the nineteenth century, but to the eighteenth.

As Thomas Gaines, one of the original architects of the Alliance and a close associate of William Lamb and Evan Jones, explained, the Alliance needed to stand in relation to the People's Party as the Jacobin Clubs of revolutionary France had stood in relation to the new democratic parliamentary government. The self-organized people of the Alliance would serve as "a mighty base of support" for Populist candidates when they legislated democratically and a strong admonishing force when they did not. By that means the people would retain control over their own movement and not surrender it to their own politicians. While this view incorporated a sophisticated understanding of the essential ingredients of mass democracy, after the destruction of the cooperatives the organizational means to keep the Alliance alive unfortunately did not exist. The Alliance founders spoke of the continuing "educational" value of the order as a grass-roots forum for political debate and they spoke of "the community" of the Alliance, sometimes in far-ranging terms that incorporated advanced ideas on women's rights and democratic human relations generally. But a community cannot persist simply because some of its members have a strong conviction that it ought to persist. A community, even one seeing itself as a "brotherhood" and "sisterhood," needs to have something fundamental to do, an organic purpose beyond "fellowship" that reaffirms the community's need to continue its collective effort. And this, after the collapse of the cooperatives, the Alliance failed to have. As the cooperatives went down in 1890–93, the Alliance organizations across the nation gradually, often rather quickly, became smaller. The order therefore could not serve either as a "mighty base" of support or as an influential force for ethical admonition within the third party movement. Even though many, most, or all of its former members were Populists in good standing, by 1893 the internal democratic cohesion of the Alliance movement had begun to weaken—in every state in America.

The era of the People's Party, therefore, may be seen as a period of gradual decline in organized democratic energy in Gilded Age politics. To be sure, there were vibrant rallies; spectacular editorials in the journalistic flagships of the Reform Press Association; moments of heady victory; and intense, colorful, and often bitter campaigns. But though all of these things were done in the name of the people's movement, it no longer, in a real sense, *was* a people's movement: the third party

had no interior mass base as its core; it had only individual
adherents who voted for it on election day.

In democratic terms, the structural weakness of the People's
Party evolved from the failure of its organizers, in the founding
convention of 1892, to understand that the third party, to be
authentically democratic, had to be organized as a mass party
with a mass membership. It was organized instead, like all large
American parties before and since, as a representative party,
with elite cadres of party regulars dominating the organizational
machinery from precinct to national convention. The People's
Party spoke, rather more tellingly than most American parties
have ever done, in the name of the people. But in structural
terms the People's Party was not made up of the people; it was
comprised of party elites. Its ultimate failure, therefore, was
conceptual—a failure on a theoretical level of democratic analysis.*

Nevertheless, despite these necessary qualifications the People's Party was, thanks to the sheer emotional and organizational
intensity that brought it into being, a political institution of
unique passion and vigor in American history. There were,
however, always significant gradations of participation in the
movement culture. For those who had learned their cooperative
lessons most thoroughly, the original Alliance dream sustained
itself throughout the entire life of the third party crusade. But
upon all those who had participated in these experiences in less
vivid ways the pull of the received culture inexorably worked its
will. And in many new Alliance states where the third party was
numerically small Populism became ideologically fragile as well.
In the state that led the farmers to the Alliance, Texas, and in
the state that led the Alliance to the People's Party, Kansas, the
agrarian dream possessed continuing, though gradually diminishing, democratic intensity. It maintained for a while an interesting thrust in Georgia, Alabama, California, and South Dakota
as well. Everywhere else the democratic vision had, by 1894,
begun to grow noticeably weaker and, in some places, had begun

* This failure is shared, of course, by the modern major parties of America—
a fact that undergirds the hierarchical corporate structure of contemporary
American society and politics.

to wither with some abruptness. Thus while the People's Party
was many things, even many democratic things, it was not an
unsteepled chapel of mass democracy—its own functionaries
saw to that, conclusively, in 1896. Beyond this, each stolen
election and act of terrorism in the South and each "practical"
coalition in the West served to chip away the morale and sap the
energy of the movement. It took the strongest kind of parallel
communities and the steadiest kind of movement leadership to
keep the democratic idea alive in the presence of the sustained
hierarchical cultural attacks (including vigilante attacks) to which
the reform movement was subjected. After the defeat of the
cooperatives, the pressure was too great for the People's Party,
alone, to bear. The movement lost its animating reality and, in
the end, became like its shadow and succumbed to hierarchical
politics.

Long before that moment of self-destruction in 1896 the
movement's organizational source, the National Farmers Alliance, having rendered its final service to farmers by creating the
new party of reform in 1892, had moved into the wings of the
agrarian revolt. In so doing, the Alliance transferred to the
political arena the broad aims it had failed to accomplish through
its cooperative crusade. The organizational boundaries of the
People's Party were fixed by the previous limits of the Alliance.
The greenback doctrines of the Alliance were imbibed by those
who participated in the cooperative crusade, but by very few
other Americans. Though several Western mining states that
were not deeply affected by Alliance organizers achieved a
measure of one-plank silverism, and Nebraska produced its
uniquely issueless shadow movement, no American state not
organized by the Alliance developed a strong Populist presence.
The fate of the parent institution and its political offspring was
inextricably linked.

The largest citizen institution of nineteenth-century America,
the National Farmers Alliance and Industrial Union persisted
through the 1890's, defending the core doctrines of greenbackism within the People's Party and keeping to the fore the dream
of a "new day for the industrial millions." Its mass roots severed
by the cooperative failure at the very moment its hopes were

carried forward by the People's Party, the Alliance passed from
view at the end of the century. Its sole material legacy was the
"Alliance warehouse" weathering in a thousand towns scattered
across the American South and West. In folklore, it came to be
remembered that the Alliance had been "a great movement"
and that it had killed itself because it had "gone into politics."
But at its zenith, it reached into forty-three states and territories
and, for a moment, changed the lives and the consciousness of
millions of Americans. As a mass democratic institution, the saga
of the Alliance is unique in American history.

In cooperative defeat and political defeat, the farmers of the
Alliance and the People's Party lost more than their movement,
they lost the community they had created. They lost more than
their battle on the money question, they lost their chance for a
measure of autonomy. They lost more than their People's Party,
they lost the hopeful, embryonic culture of generosity that their
party represented. The stakes, as the Alliance founders had
always known, were high, for the agrarian dream was truly a
large one—a democratic society grounded in mass dignity.

13

What is to be concluded of Populism's three opponents—the
shadow movement, the Democratic Party, and the Republican
Party?

The sometimes formless, sometimes silverite expression which
historically has been regarded as "American Populism" and
which has appeared in these pages as the "shadow movement"
may be judged to be trivial as a political faith. A natural
outgrowth of the intellectual and programmatic aimlessness of
the Northwestern Farmers Alliance, the shadow movement had
historical meaning only in the context of the actual agrarian
movement mounted by the National Farmers Alliance and
Industrial Union. Devoid of genuine political substance, the
shadow movement was patently less interesting than Populism,
being little more than an imitative reprise on the superficial
agitation over the "Crime of '73" that had helped to sidetrack

greenback reform a generation earlier.* The second silver drive proved no more coherent than the first. The shadow movement, as it triumphed in 1896 in Kansas, in 1894 in North Carolina, and in 1890 in Nebraska, had common characteristics: it was led by politicians, its tactics were defined by them, and the usefulness of such tactics depended upon the successful avoidance of the intellectual content embedded in the Populist platform. Some fusionist politicians, like Senator Peffer of Kansas, gave up the Populist faith reluctantly and only temporarily; others, like Allen of Nebraska, had never known it. But in embracing the silver expedient, they all identified themselves with a cause that originated outside the People's Party and was financed by mining interests whose goals were inimical to the objectives of Populism.

The symbolic spokesman of both the Northwestern Farmers Alliance and the shadow movement was Jay Burrows of Nebraska. He, having never grasped the meaning of the cooperative crusade and never having experienced its political lessons, could write in 1896 that William Jennings Bryan was "practically a Populist except in name."

Organizationally derivative, intellectually without purpose, the shadow movement never fashioned its own political premise; inevitably, its sole criterion became conventional politics itself— the desire of office-seekers to prevail at the next election. In attempting to address their own short-run electoral needs, the tacticians of fusion adopted as their own the political purposes of the silver mineowners. But having no political or cultural base independent of others, the fusionists had no hope of surviving on the goals of others. Their moment of tactical triumph, lasting a few hours at the Populist convention in 1896, was inherently ironic: in achieving the nomination of William Jennings Bryan, they not only ensured their own organizational extinction, they also provided a way for farmers to submerge themselves in the new two-party corporate politics of modern America.

The silver crusade nevertheless possessed considerable cul-

* See Chapter 1.

tural significance for what it revealed of the state of political
consciousness of those millions of Americans who did not
respond to the Omaha Platform but who did embrace the free
coinage panacea. The silver Democratic Party of William Jennings Bryan, like the Gold Democracy of Grover Cleveland and
the "sound money" Republican Party of William McKinley, was,
intellectually, an anachronism. As negative reactions to the
Omaha Platform, the politics of silver and the politics of gold
illustrated the narrowing political range of the corporate society
that was then emerging. In monetary analysis, Gold Democrats
were neither better nor worse than Silver Democrats, merely
their equals in hyperbole and in their misunderstanding of how
monetary systems worked. Both represented the triumph of
form over substance.

In statesmanship, the Bryanites, not achieving office, left no
record; the Cleveland Democracy, however, repeatedly proved
its inability to perceive the world around it and demonstrated
rather forcefully its resulting incapacity to govern. Frozen in a
philosophy of laissez faire, Cleveland was virtually immobilized
by the depression of the 1890's, and was unable to respond
either to the needs of the unemployed or to the imperatives of
the inevitable financial crisis. The ultimate effect of his frantic
gold bond issues was to transfer sizable portions of the government's reserve from the United States Treasury to the Eastern
financial community. All was done in the name of "sound
money."

On the other hand, the Republican Party, an increasingly
narrow-gauged but powerful engine, made a determined show
of running both the nation and its economy. Thanks to the
energy of the American industrial system, a demanding and
ruthless contrivance, but, relatively speaking, an efficient one,
the Republicans were able to do just that. They liked the
contrivance, even celebrated it with a self-serving and narrow
insularity that blinded them to the costs it exacted in both human
and natural resources, in the beauty of the land itself, and in the
vigor and range of the national culture. But they did all these
things with a public display of confidence and were able to
escape relatively unscathed. The Republicans, it should be noted,

did not respond to the depression with any more verve than Cleveland had managed. But with its newly consolidated majority position, the Republican Party—except for the Wilsonian inter-lude—ruled the nation for two generations. During that period America added to its material wealth because of the absence of feudal restraints upon the imaginativeness of its people, because of the energy of both its entrepreneurs and their employees, and because of the great bounty nature had bestowed upon the American continent. The political institutions of the nation were vastly overshadowed in importance by these deeper rhythms that propelled an unbalanced and provincial democracy toward the crises that awaited it in the twentieth century.

14

When those crises came, twentieth-century America found itself still firmly bound to the nineteenth-century orthodoxies against which the Populists had rebelled—presumptions that outlined the proper limits of democratic debate, the prerogatives of corporate money in shaping political decision-making, and the meaning of progress itself. Though the first years of the new era were no longer called the "Age of Excess"—they were, in fact, called the "Progressive era"—the acceptable political bound-aries for participants in both parties received their definition from the economic values that triumphed politically in the Gilded Age. When the long Republican reign came to an end in 1932, the alternatives envisioned by the Democrats of the New Deal unconsciously reflected the shrunken vistas that remained culturally permissible. Aspirations for financial reform on a scale imagined by greenbackers had expired, even among those who thought of themselves as reformers. Inevitably, such reformers had lost the possibility of understanding how the system worked. Structural reform of American banking no longer existed as an issue in America. The ultimate cultural victory being not merely to win an argument but to remove the subject from the agenda of future contention, the consolidation of values that so successfully submerged the "financial question" beyond the purview of succeeding generations was self-sustain-ing and largely invisible.

The complacency of the nation's intellectual elite of the Gilded Age, it turned out, ranged beyond the dull monometallism of academic economists such as J. Laurence Laughlin or the superficial satire of E. L. Godkin; it enveloped even those enlightened literary arbiters who considered themselves in the vanguard of the new realism. When a sympathetic William Dean Howells could unwittingly and patronizingly describe the farmers of the Alliance as "grim, sordid, pathetic, ferocious figures" and characterize the cooperative crusade, about which he almost certainly knew little or nothing, as "blind groping for fairer conditions," the cultural barriers to analytical clarity were manifestly settling into hardening concrete. In the new century, gifted historians such as Richard Hofstadter not only followed in the Howells tradition, their interpretations of the American experience captured the cultural high ground within the American academy. In the prevailing view, "the system," beset with human frailty as it was, "worked." While it had flaws, more often than not they materialized from the irrationality and spite of the general populace rather than from any failings on the part of the nation's commercial and intellectual elite. This thought, of course, has proved enormously consoling to the elites, and accounts for the undiminished emotional appeal of this reigning intellectual tradition.*

* It is interesting to observe that one group of Americans, the nation's leading novelists, have consistently resisted both the language of genteel apologia and the language of celebration that combine to inform the mainstream academic view of the American past. The contrast between the sanguine, culturally confined literature of American history and the brooding chronicle of endurance and tragedy that characterizes the nation's best fiction is an interesting contemporary phenomenon. The American historical tradition of conveying the national experience as a purposeful and generally progressive saga, almost divinely exonerated, as it were, from the vicissitudes elsewhere afflicting the human condition, may perhaps account for the remarkably different levels of impact American historians and American novelists have had on non-Americans. While a number of American novelists, from Melville to Faulkner, are admired throughout the world, no American historian has remotely achieved similar standing.

It may be argued that the historical approaches that are traditional within the American academic mainstream have failed to uncover mass striving and mass defeat because, on the deepest psychological level, more specific conceptual and analytical tools were not thought to be needed to describe social realities in this "democratic" nation. It was simpler to dismiss moments of unseemly discord as,

Grounded in such a view of the national past, the political sensibility that materialized within the cramped American party system of the twentieth century found itself too rigid and too dangerously condescending to support truly expansive vistas of democratic possibility, such as those afforded by the continuing aspirations of the millions of "the people" who, because of manners, occupation, or skin color, could not gain access to the benefits of the progressive society. Within the confines of this narrowing social outlook, Americans seem to have lost the capacity to think seriously about the structure of their own society. Words like "inevitability," "efficiency," and "modernization" are passively accepted as the operative explanations for the increasingly hierarchical nature of contemporary life.

Three central issues of Populism illustrate the mass resignation characteristic of modern society and reveal the extent Americans have lost contact with the democratic premises of earlier eras. Those issues concern, first, land ownership in America; second, the hierarchical nature of the nation's basic financial structure; and third, the consuming threat that corporate centralization poses to the democratic heritage itself. In much the same way that centralization of land was characteristic of feudalism, it has become increasingly characteristic of modern America. The centralization of land ownership in American agriculture has for years been fashionably attributed to the competitive "efficiency" of "economies of scale" by large-unit corporate farming interests—as compared to the presumed limitations in the operations of family farms. Awaiting agricultural analysis, em-

in Howells's phrase, "blind groping for fairer conditions," or, in Hofstadter's, as the irrationalities of people who saw themselves as "innocent pastoral victims of conspiracies hatched in the distance." Whether the general failure to focus upon the depth and permanence of hierarchical structures of power and privilege in America is a cause or a result of inherited modes of cultural narrowness is, perhaps, debatable, but the complacent cast of the historical genre scarcely is. Though there are noteworthy exceptions, the standard American history texts used in high schools and colleges throughout the nation are impressive examples of scholarly deference to the received culture. As such they serve to reveal how thoroughly the values of the corporate state suffuse contemporary intellectual life in the United States. The nation's most renowned novelists, of course, have taken pains to position themselves securely outside this narrow and uncreative orthodoxy.

barrassingly enough, is a considerable amount of evidence suggesting that the real economies of scale are not technical, but artificial, produced by the monopolistic practices of suppliers and purchasers and further undergirded by federal subsidy and tax policies. But while the concept of "economies of scale" remains debatable, the more germane historical reality is that centralization of American land was well advanced even before corporate agriculture could prove or disprove its "efficiency." It was simply a matter of capital and the power of those having capital to prevent remedial democratic legislation. The failure to provide short-run credit for seventy years would seem to be the operative ingredient in these dynamics which has been rather overlooked.

The one agricultural adventure of the New Deal that most nearly approached Charles Macune's objective of benefiting "the whole class" was the resettlement program of the Farm Security Administration, designed to provide land loans at low interest to enable black and white tenants to become owners of family farms. The program was initiated on a small scale in isolated pockets in the South and West in the 1930's. Though the program was successful, land acquisition was halted in 1943. In that year the program, long under attack from large-unit farming interests within the Department of Agriculture, from large-unit farming interests in the Farm Bureau Federation, and from congressmen responsive to the lobbying power of large-unit farming interests, was killed. The process of land centralization across America has since accelerated, especially affecting black landowners in the South.

In the absence of significant literature on the subject, land centralization is a process that remains obscure to most Americans, but one they may feel no right to inquire into—given the fact that land centralization is sanctioned by the culture itself. Indeed, the remarkable cultural hegemony prevailing militates against serious inquiry into the underlying economic health of American society, so this information is, first, not available, and, second, its non-availability is not a subject of public debate. Large-scale property ownership in America is a legal secret secluded in "trusts," "street names," and "nominees" beyond the reach of any democratic institution in the society.* Though Populists had the self-possession to struggle against land syndicates, Americans no longer contest the matter.

Similarly, the nation's hierarchical banking system dominates the lives of modern Americans in subtle and pervasive ways. For millions of Americans one of the central purposes of long years of striving is the desire to own a home of their own. The pattern of interest rates, together with federal housing programs tailored to the self-interest of the banking community, have effectively placed such home ownership beyond the reach of the majority of working class Americans. As informed political scientists know, and have reported, of all Washington lobbies, the most powerful is the banking lobby. Even the "Pentagon lobby" must shape its needs to conform to the imperatives of the commercial bankers. Yet the idea of a democratic monetary system—the operative dynamic of American Populism—is simply not something that Americans seem any longer to aspire to. It is not "practical" to have such large democratic goals. Thus does modern sophistication serve as a defense for modern resignation.

Inexorably, the consolidation of economic power in corporate America has shaped an entirely new political landscape, one in which the agenda of possible democratic actions has shrunk significantly. The Populist fear that corporate concentration would undermine the popular autonomy necessary to the preservation of authentic democratic dialogue has been realized. Modern politics takes place wholly within the narrowed boundaries of the corporate state. In most circles, it is now considered bad manners to venture outside these boundaries. While most Americans do not venture, they also do not celebrate the limits. They cannot, however, find a culturally sanctioned way to express their anxiety politically. A heartfelt but unfocused

* Informed research in this area has not yet begun—and cannot be—until the disclosure laws are reformed. See U.S. Congress, Senate Committee on Government Operations, *Disclosure of Corporate Ownership,* Staff Report of the Subcommittee on Budgeting, Management and Expenditure, 93rd Cong., 2nd sess., May 4, 1974 (Washington, D.C., 1974); U.S. Congress, House Committee on Banking and Currency, *Commercial Banks and Their Trust Activities, Emerging Influence on the American Economy,* Staff Report for the Subcommittee on Domestic Finance, 90th Cong., 2nd sess., July 8, 1968 (Washington, D.C., 1968).

discontent about "politics" has therefore become a centerpiece of the popular subculture across the nation.

Twentieth-century people, reduced to the sobering knowledge that theirs, politically, is the least creative century of the last three, take refuge in private modes of escape and expression, found largely through the pursuit and consumption of products. The corresponding decline in the vitality of public life verifies the constraints of modern political thought. "The people," though full of anxiety, do not know what to do politically to make their society less authoritarian. Language is the instrument of thought, and it has proven difficult for people to think about democracy while employing hierarchical terminology. On the available evidence, twentieth-century people around the globe are paying a high price for their submission to the hierarchical languages of political analysis that have grown out of the visions of Adam Smith and Karl Marx. The problem that will doubtless interest future historians is not so much the presence, in the twentieth century, of mass political alienation, but the passivity with which the citizenry accepted that condition. It may well become known as the century of sophisticated deference.

As objects of study, the Populists themselves were to fall victim to the inability of twentieth-century humanists of various ideological persuasions to conceive that authentic political substance might originate outside such acceptable intellectual sources as the progressive, capitalist, middle classes or the European socialist heritage. Since both traditions continued to employ encrusted political terms rooted in outdated nineteenth-century languages of prophecy, the antiquarian nature of modern politics became difficult to overstate. From all quarters, Populists were denied authentic historical association with their own movement. Accordingly, their chief legacy—a capacity to have significant democratic aspirations, a simple matter of scale of thought— faded from American political culture. The reform tradition that materialized in the twentieth century was intimidated and, therefore, unimaginative. Harry Tracy and other Populist theoreticians had called for 300 per cent inflation through a return to the per capita circulation of 1865, on the sole and valid ground that the sub-treasury system was intelligent and would provide an immediate remedy for a central and long-neglected flaw at the heart of American capitalism. The capacity to think politically on such a scale was Populistic; it passed from the mainstream American reform tradition with the defeat of Populism.

By this process, the relatively expansive pre-industrial sensibilities that had animated Thomas Jefferson, George Mason, and the original Anti-Federalists gradually lost that strand of democratic continuity and legitimacy which, in fact, connected their time and their possibility to our own through the actions of Americans who lived in the interim: the Populist connecting link was lost to the heritage.* The egalitarian current that was part of the nation's wellspring became not a constantly active source of ideas, but a curious backwater, eddying somewhere outside both the conveyed historical heritage and the mainstream of modern political thought that necessarily builds upon that heritage.

* Unlike Jefferson, of course, Populists were not dedicated to small government because they believed such an entity could not cope with the power of concentrated capital. Rather, Populists sought democratic government, as Jefferson himself had.

Populist theory poses the central twentieth-century political question: can large government be democratic? The history of twentieth-century industrial societies indicates not—at least not within the prevailing conceptual limitations of traditional capitalism and traditional socialism. Unfortunately, the idea that workable small-unit democracy is possible within large-unit systems of economic production is alien to the shared presumptions of "progress" that unite capitalists and communists in a religious brotherhood. So much so that the very thought tends to give people a headache. Intellectual short-circuits crackle everywhere. The topic, therefore, is not one the young are encouraged to speculate about; such a possibility challenges our settled resignation and puts people ill at ease. It is simpler to sustain one's morale by teaching the young not to aspire too grandly for too much democracy.

The conclusion is transparent: the intellectual range of modern industrial societies is quite narrow. One observes that this conclusion is avoided by participants in the mainstream of capitalist and socialist societies because, to do otherwise, in sophisticated circles, is not career oriented and, in unsophisticated circles, is unpatriotic. On the historical evidence, it seems possible to conclude that Mark Hanna and Henry Vincent (for different reasons, of course, since Hanna was not a democrat) would have understood these dynamics without too much psychic strain. On the other hand, it seems probable that William Jennings Bryan, in his provincial innocence, would not have understood. Bryan may be seen, therefore, to be one of the participants in Gilded Age public life who most nearly resembles "modern" political figures.

The result is self-insulation: the popular aspirations of the people of the "third world" in the twentieth century have easily become as threatening to modern Americans as the revolt of their own farmers was to goldbugs eighty years ago. Though American foreign policy and American weapons have defended anachronistic feudal and military hierarchies in South America, Africa, and Asia, such actions being justified at home as necessary to the defense of "democracy," neither the policy nor the justification has proved notably persuasive to the non-Americans who are the mass victims of such hierarchies. The resulting unpopularity of America puzzles Americans. The policies them-selves, however, are not debatable within the limits of public dialogue sanctioned in modern America. Under such constraints, the ultimate political price that Americans may be forced to pay for their narrowed cultural range in the twentieth century has emerged as a question of sobering dimension.

15

However they were subsequently characterized, Populists in their own time derived their most incisive power from the simple fact that they declined to participate adequately in a central element of the emerging American faith. In an age of progress and forward motion, they had come to suspect that Horatio Alger was not real. In due course, they came to possess a cultural flaw that armed them with considerable critical power. Heretics in a land of true believers and recent converts, they saw the coming society and they did not like it. It was perhaps inevitable that since they lost their struggle to deflect that society from its determined path of corporate celebration, they were among the last of the heretics. Once defeated, they lost what cultural autonomy they had amassed and surrendered their progeny to the training camps of the conquering army. All Americans, including the children of Populists, were exposed to the new dogmas of progress confidently conveyed in the public school system and in the nation's history texts. As the twentieth-century recipients of this instruction, we have found it difficult to listen with sustained attention to the words of those who dissented at the moment a transcendent cultural norm was being fashioned.

In their own era, the agrarian spokesmen who talked of the "coming revolution" turned out to be much too hopeful. Though in the months of Populist collapse and for successive decades thereafter prosperity eluded those the reformers called the "producing classes," the growing industrial society preserved the narrowed boundaries of political dialogue substantially intact, as roughly one-third of America's urban workers moved slowly into the middle class. The mystique of progress itself helped to hold in muted resignation the millions who continued in poverty and other millions who, for reasons of the exclusiveness and white supremacy of the progressive society, were not permitted to live their lives in dignity.

As the first beneficiary of the cultural consolidation of the 1890's, the new Republican orthodoxy, grounded in the revolutionary (and decidedly anti-Jeffersonian) political methods of Mark Hanna, provided the mores for the twentieth century without ever having to endure a serious debate about the possibility of structural change in the American forms of finance capitalism. Political conservatives nevertheless endured intermittent periods of extreme nervousness—such as was produced in 1933 by the nation's sudden and forced departure from the gold standard. Given the presumed centrality of a metallic currency, it took a while for cultural traditionalists, including bankers, to realize that the influence of the banking community had not suffered organic disturbances—J. Laurence Laughlin to the contrary notwithstanding. Though the pattern of interest rates during and after World War II continued to transfer measurable portions of the national income from both business and labor to bankers—in the process burdening the structure of prices with an added increment of cost as well as changing the very structure of industrial capitalism—disputes over the distribution of income within the whole society did not precipitate serious social contentions as long as America maintained a favorable international trade and investment balance. It remained clear, however, that unresolved questions about the inherited financial system might well make a sudden and unexpected reappearance if, at any time in the second half of the twentieth century, shifts in world trade and the cost of imported raw materials placed severe forms of competitive pressure on the American economy and on the international monetary system. At such a moment the cultural consolidation fashioned in the Gilded Age would undergo its first sustained re-evaluation, as the "financial question" once again intruded into the nation's politics and the issues of Populism again penetrated the American consciousness. That time, while pending, has not yet come.

For their part, Gilded Age traditionalists did not view the conclusive triumph of the corporate ethos as a foregone conclusion. Themselves insecure in an era of real and apparent change, they were unable to distinguish between authentic signs of economic dislocation and the political threat represented by those who called attention to those signs. On this rather primitive level the politics of the era resolved itself, and the progressive society was born. As an outgrowth of its insularity and complacency, industrializing America wanted uncritical voices of celebration. The agrarian radicals instead delivered the warning that all was not well with the democracy. They were not thanked.

Today, the values and the sheer power of corporate America pinch in the horizons of millions of obsequious corporate employees, tower over every American legislature, state and national, determine the modes and style of mass communications and mass education, fashion American foreign policy around the globe, and shape the rules of the American political process itself. Self-evidently, corporate values define modern American culture.

It was the corporate state that the People's Party attempted to bring under democratic control.

Afterword

What became of the radicals—the earnest advocates who spawned the agrarian revolt? The subsequent careers of the people who have filled these pages merely testified to the diversity of American life. Some Populists adjusted to the collapse of their movement with what others in the ranks regarded as entirely too much poise and equanimity. Others looked desperately for a new political home, and a few, finding none and unable to bear the consequences, committed suicide.

In Kansas, Senator William Peffer recoiled from his brief sojourn into fusion in 1896, publicly described it as a decisive error, and was defeated for re-election. He dallied briefly as a third party Prohibition candidate and thereafter returned to the Republican Party. The fusion-minded chairman of the Kansas party, John Breidenthal, whose political career extended back through the Union Labor Party to the original Greenback Party, flirted with socialism and then grudgingly became a Republican. One of his mid-road opponents in Kansas, G. C. Clemens, became the Social Democratic candidate for Governor in 1900. At Clemens's opening campaign meeting, he was introduced by another convert to socialism, Lorenzo Lewelling, the ex-Governor and former ardent fusionist. Jerry Simpson went to New Mexico, where he became a land agent for the Santa Fe Railroad.

Frank Doster, the socialist judge, became the Populist Chief Justice of the Kansas Supreme Court after being attacked as a

"shabby, wild-eyed, rattle-brained fanatic" by William Allen
White, the Emporia editor. Doster's evenhanded decisions and
eloquent written opinions won the grudging respect and, eventually, the freely acknowledged admiration of the Kansas bar.
In the new century, the courtly jurist had the satisfaction of
watching the slow evolution of William Allen White into a
Progressive, gracefully accepted White's public apologies, and
remained an unreconstructed critic of many features of American capitalism throughout his long life. Doster worked actively
for woman's suffrage in Kansas, considered the election of
Franklin D. Roosevelt a hopeful sign, and died in 1933 while
drafting agrarian reform legislation to help the increasing
number of tenant farmers in Kansas.

Julius Wayland developed the *Appeal to Reason* into the nation's
foremost socialist newspaper. One famed special edition of a
million copies required the labor of most of the citizens of
Girard, Kansas, to mail. Though a successful businessman as
well as an ardent socialist, Wayland ultimately lapsed into
despondency and committed suicide.

Marion Butler became a greenbacker and William Allen
became a mid-roader—at least fleetingly. After Allen's retire-
ment from the Senate, he ultimately decided the policy of fusion
had been a mistake. He attended one of the final, desperate
"reorganizing" meetings of mid-road Populists in Denver in
1903 and then focused on his law practice. Marion Butler
declared his conversion to a fiat currency in one of his last public
addresses in the United States Senate. At the time, 1901, Butler
knew he was a lame-duck Senator because Populist-Republican
fusion in North Carolina had fallen victim to Democratic terrorism in 1898. In a massive campaign of white supremacy,
punctuated by gunfire and arson, "the party of the fathers" had
been swept back into office that year. During Populism's last
stand in North Carolina in 1900, two of Butler's old opponents,
Harry Tracy and "Stump" Ashby, came into the state in a vain
effort to help him combat the organized politics of white
supremacy. Butler was at that time one of the last surviving
Populist officeholders in the South, and his conversion to the
greenback cause apparently was the fruit of his belated associ-

ation with Tracy, the third party's foremost monetary theorist. Butler's break with the Democratic Party, which followed hard on the heels of the lessons he learned in the fusion campaign of 1896, was permanent. He gradually became more liberal on the race issue. Butler passed into the Southern Republican Party in the new century and never thereafter came to terms with the party of the fathers.

Tom Watson did. The frustration of his Populist years dogged Watson through a hopeless 1908 campaign in which he served as the last presidential candidate of the diehard Populist remnant. After more than a decade of stolen elections and what he regarded as fusionist betrayals, Watson became deeply embittered. He eventually blamed blacks, Catholics, and Jews for his own, and the nation's, political difficulties. He became a violently outspoken white supremacist, anti-Semite, and defender of the Republic against the papal menace. In the twilight of a life steeped in personal tragedy and blunted dreams, and consumed at its end by the political malice he had developed as a battered campaigner, Watson in 1920 won a surprising victory and became United States Senator from Georgia. He died in office in 1922. One other prominent Southern Populist, James "Cyclone" Davis, had an almost identical political career. Alienated from most of his fellow Populists in Texas because he was not sufficiently alienated from the Democratic Party, Davis, a covert Texas fusionist in 1896 and an overt one in 1898, made the full return to "the party of the fathers" as the new century opened. He became a prohibitionist and white supremacist, worked for the Ku Klux Klan, and won election for one term as a Congressman in 1916, taking the old seat that had been denied him by "Harrison County methods" in 1894. As late as 1939, the white-bearded Davis could be heard on the streetcorners of Dallas, making flamboyant speeches on the need to control Wall Street and the necessity of white supremacy.

Reuben Kolb in Alabama and James Weaver in Iowa both remained Bryan Democrats after 1896. Weaver became the Mayor of Colfax, Iowa, and died in that office in 1912. Minnesota's Ignatius Donnelly enjoyed a minor career as a literary figure. Mann Page, the Virginia Populist, became one of the last presidents of the National Farmers Alliance and gradually and reluctantly returned to the Democratic Party of Virginia. The editorial spokesman of Virginia Populism, Charles Pierson, the Oxfordian who edited the Virginia *Sun,* became a Debsian socialist.

Many of the most renowed radical organizers of the Alliance, relatively young men in the early cooperative days in the 1880's, lived well into the twentieth century. South Dakota's Henry Loucks was still writing antimonopoly pamphlets that extolled the virtues of farmer cooperatives as late as 1919. J. F. Willits, leader of the Kansas Alliance, became a socialist. W. Scott Morgan of Arkansas penned a brooding attack on the racial demagoguery of the Southern Democracy and saw it published as a novel in 1904. (It must be said that Morgan's political instincts surpassed his literary gifts.) Joseph Manning, Alabama's relentlessly energetic Populist, fought fusion to the bitter end in 1896 and then joined the rapidly shrinking Alabama Republican Party in the midst of the politics of disfranchisement. He eventually moved to New York, where on one occasion he was honored by a black organization with whom he shared progressive political sympathies. Like Morgan of Arkansas, Manning tried to energize other white Americans to do something about the party of white supremacy in the South. If Manning's politics coalesced easily with Morgan's, his book, *The Fadeout of Populism,* published in 1922, unfortunately also confirmed him as Morgan's equal as a writer.

Henry Vincent spent his life as a printer, editor, and reformer. In the 1920's he received the support of a number of American intellectuals and progressives, including John Dewey and Arthur Garfield Hays, in a new venture called *The Liberal Magazine.* His brother Leopold, whose lyrics found their way into the *Alliance Songster* during the hopeful days of the Alliance national organizing campaign in 1890–91, married a young woman of progressive views and settled in Oklahoma. "Stump" Ashby, the "famous agitator and humorist" and perhaps Populism's most eligible bachelor, also migrated to Oklahoma. He married a daughter of the Comanche Indian nation, and sired a large family. Along with his old radical colleague, S. O. Daws, the original "traveling lecturer" of the Alliance, Ashby helped organize the Oklahoma Farmers Union in the new century. A rangy patriarch of the left whose oratorical powers and humor brought him friends wherever he went, Ashby died in Octavia, Oklahoma, in 1923.

William Lamb became one of the most prosperous of all the old radical organizers. Much of the land he acquired for his Montague County farm became the site of the town of Bowie. Though the early Montague cooperative gin and mill that he and thirteen other Alliancemen helped underwrite in 1886 collapsed because of lack of access to credit, eventually Lamb was able to build a similar enterprise out of the proceeds of the sale of town lots. "Lamb and Hulme, millers and ginners," established a modified "sub-treasury" warehouse of their own by permitting farmers to store their cotton without charge while awaiting higher prices. The firm did one of the largest businesses in the North Texas farming country until it was destroyed by fire in the middle 1890's. In 1906, Lamb's hybrid fruits won prizes at the Fort Worth Exposition—though the aging boycotter did not quite fulfill the role of a gentleman farmer. Among other things, he remained politically radical. Many small American towns before World War II could claim one old iconoclast whose political views had been shaped by bygone struggles. In Topeka, Kansas, his name was Frank Doster; in Bowie, Texas, his name was William Lamb.

2

But if the post-Populist careers of Henry Vincent, "Stump" Ashby, Joseph Manning, and Henry Loucks, among many others, testify to the endurance of the Populist vision, and if the twentieth-century careers of Tom Watson and "Cyclone" Davis illustrate the vulnerability of that vision, the post-Populist career of one other agrarian spokesman points to something else about the American experience of even broader implication—the extent to which the American ideas of progress and democracy have been, and remain today, deeply disfigured by the enduring stain of white supremacy.

The most famous black orator in all of Populism was the third party's state executive committeeman from Texas, John Rayner. A schoolteacher in East Texas, Rayner organized a corps of assistants to work for Populism in black districts of the state. As signs of progress appeared and Rayner's reputation grew among third party men, he became a leader of the Texas People's Party and ranked as an orator only behind two nationally known third party men, James Davis and H. S. P. Ashby.

With the collapse of the third party in 1896, and the advent of black disfranchisement after the turn of the century, Rayner faced long years of humiliation as a well-known public man whose personal politics had become untenable in the changed world of the new century. An organizer who had braved sniper fire as well as physical assaults in the cause of Populism, Rayner relinquished one by one the goals of the program of interracial political action he had carried in the 1890's. By 1904, after repeated rebuffs, he had been reduced to one goal, the cause of private Negro education, which could only be supported by wealthy white people in Texas. Gradually, desperately, Rayner learned the language that would open this channel of communication. It was a language of submission that at times neared, if it did not breach, the boundaries of personal abasement. With these credentials, he raised funds for the "Farmers Improvement Society" and its agricultural school to keep Negroes down on the farm. By 1910 Rayner could draw what satisfaction was possible from the knowledge that he represented "the right kind of Negro thought" in the opinion of those who shared in power in Texas.

He returned to the Republican Party, and his old-time fire occasionally reappeared in angry letters about the racism of the "Lily Whites." In 1912, the ground for a free-thinking black to stand upon in the South had narrowed to the vanishing point, and Rayner accepted private humiliation rather than endure the psychic cost of further ritual servility. He went to work for John Kirby, a famous Texas lumber king, and spent the remaining years of his life as a labor agent recruiting Mexican peasants and transporting them to East Texas to work in Kirby's lumber mills. Rayner, a man who had spent his youth learning

the art of spoken and written persuasion, employing both in
political radicalism, and who in his middle years had acquired
the new art of spoken and written dissimulation, became in his
old age an instrument of exploitation of those having even
narrower options than himself. According to his descendants,
he died "very bitter." Rayner's private papers, which extend
over the period 1904 to 1916, provide a harsh insight into the
years following slavery in which an aggressive white supremacy
triumphed politically in the United States and gave rise to the
settled caste system of the first half of the twentieth century.
Before his death in 1918, Rayner pronounced a one-sentence
judgment on the political system that had defeated him throughout his life. "The South," he said, "loves the Democratic Party
more than it does God." However, it was not just the Democratic
Party that defeated him, it was the culture itself.

                                   3

Irony is the handmaiden of American radicalism, and the
sharpest ironies were reserved for Charles W. Macune. As the
Southern Mercury observed at the time, Macune, having no heart
for a radical third party, simply withdrew from the ranks of the
reformers in 1892. He never returned. He lived in the East for
a while and eventually went back to Waco, the scene of his first
great triumph in 1887 and his tactically decisive defeat at the
hands of William Lamb in 1891. He lived out his life in Texas
as a Methodist pastor, aided by his son in his final endeavor—
ministering to the agrarian poor. In 1920, Macune deposited
his reminiscences of the Alliance years in the University of Texas
library in Austin. The fifty-nine page manuscript, as enigmatic
as Macune's own career, raised more questions than it answered.
He understood the nation's economy better than most Gilded
Age economists, and he understood the limits of the cooperative
movement better than other Alliance leaders. But he never
understood the radical political world of which he was a part,
or how he lost influence in the organization he had done so
much to build. Yet, more broadly than anyone else, he lived the
entire range of the massive agrarian attempt at self-help and

experienced in a most personal way the traumatic implications of the political movement that grew out of that effort.

In his own time, Macune's sub-treasury system was attacked relentlessly, generally without intelligence, and almost always without grace. His patient explanations were rarely printed and almost never given a fair hearing. Among historians, Macune's political traditionalism, his economic radicalism, and his recurrent opportunism have combined to leave him with few admirers. His cooperative methods have won a too-easy condemnation from those who have not confronted the realities of the financial system that both energized and defeated the cooperative movement in America.

By 1889, those who understood the options best—and Macune certainly more clearly than anyone else—knew that the Alliance dream of a national federation of regional cooperatives was untenable, because of the power and hostility of the American financial community. The sub-treasury promised to save the situation, and it did so as long as it remained merely the source of economic possibility rather than the inspiration for a radical third party. The operative life of the National Alliance, on its own terms, extended over a period of less than five years, from the announcement of the cooperative goal in the spring of 1887, through the shift to reliance upon a government-supported sub-treasury system in 1890, to the cooperative failures across the nation in 1890–92, and to its final service to farmers—as a tactical aid in the creation of the People's Party. This core economic experience of the National Alliance was Macune's experience more than any other man's.

Macune's own political weakness was the general weakness of the Southern Democracy and reflected the environment in which both he and his conservative rivals matured. He had some narrow horizons. Though his ambition for himself and the Alliance encouraged him to lofty nationwide organizing strategies, he never understood the political drives of urban workers, Western farmers, or black Americans. The men he gathered around him in Washington—Terrell, Tracy, Sledge, Tillman, and Turner—were all Southerners, three from Texas, one from Tennessee, and one from Georgia. Obsessed by the challenge of freeing the Southern farmer from the crop lien, Macune performed with unusual creativity. But he did not habitually think in political terms beyond those of his own immediate environment, nor, as his actions in 1892 revealed, did he have the political courage of his economic convictions. His sectional loyalty to the party of white supremacy stained the meaning of much of his work.

Yet, having said this, it is proper to add that Macune's sub-treasury system for the "whole class" was one of the boldest and most imaginative economic ideas suggested in nineteenth-century America. More significantly for the long run, the sub-treasury plan rested upon a broad theoretical foundation regarding the use of the nation's resources for the benefit of the entire society. Macune's plan was not a completely flawless solution to the rigidity of a metallic currency, but it not only was workable with simple modifications, it was clearly superior to the rigid doctrines of either goldbugs or silverites.

In the words of a modern monetary specialist, the sub-treasury plan was "a very subtle mechanism" which "simultaneously would have contended with the problems of financing cooperatives, the seasonal volatility of basic commodity prices, the scarcity of banking offices in rural areas, the lack of a 'lender of last resort' for agriculture, inefficient storage and cross-shipping, the downward stickiness of prices paid by farmers *vis-à-vis* prices received for crops, and the effects of the secular deflation on farmers' debt burdens, all of which, in a far less comprehensive fashion, were the objects of legislation in the next five decades. . . . It would have achieved what its supporters claimed—real income redistribution in favor of 'the producing classes.' "*

Indeed, the substance of the plan, beyond its conceptual basis, was its practical response to the financial realities of an industrializing state and the new power relationships between bankers and non-bankers that those realities enshrined. On this highly relevant topic, Macune was less rigid, or less culturally confined, than almost all Gilded Age politicians and an overwhelming majority of the nation's academic economists.

Whether in behalf of working Americans seeking a home of

* See "An Economic Appraisal of the Sub-Treasury Plan," by William P. Yohe, in Goodwyn, *Democratic Promise*, pp. 571–81.

their own, or in the context of the health and flexibility of the national economy itself, the country has never found a way to confront the enormous implications of financial relationships structurally geared to the self-interest of private, commercial bankers. It is only just to concede to Charles Macune the significance of his aims, while detailing his failings. He was the boldest single theorist of the agrarian revolt. He was also, in economic terms, one of the most creative public men of Gilded Age America. But he was also a self-created victim of an awesome political irony: while he was tactically opposed to the political activists among the Alliance founders led by Lamb, Daws, Jones, and Ashby, he shared their theoretical interpretation of the ills of the American version of capitalism. Indeed, Macune's report on the monetary system at St. Louis in 1889 specifically and repeatedly denounced the power that bankers exercised as a "class." Macune's tactical difficulty was that, in the deepest psychological dimension of his personal autonomy, he could not bear where his economic analysis carried him in political and social terms. Macune would fit well, it may be seen, into the culturally confined politics of the twentieth century.

Yet in spite of everything that befell him, including that which he brought on himself, Charles Macune was representative of an important part of the spirit of the agrarian revolt. He doubtless agreed with the prophetic warning delivered by S. O. Daws, the father of the Alliance movement and the earliest spokesman for the democratic vision that was to flower into Populism. On the eve of the first great lecturing campaign to organize the South in 1887, Daws wrote: "If the Alliance is destroyed, it will be some time before the people have confidence in themselves, and one another, to revive it, or organize anything new." This Populistic prophesy, correct as it has proven to be, is part of the legacy of the agrarian democrats to their descendants in the twentieth century.

A Critical Essay
on Authorities

Since the National Farmers Alliance and the People's Party were sequential expressions of the same popular movement and the same democratic culture, the gradual evolution of the cooperative crusade that generated both was the central component of the agrarian revolt. This understanding came largely from primary sources: early Alliance newspapers such as the *Rural Citizen* (Jacksboro), *Southern Mercury, American Nonconformist, Kansas Farmer, Progressive Farmer, The Advocate,* and the *National Economist;* later, the journals of the reform press association throughout the South and West, together with the surviving private papers, organizing pamphlets, and books of such agrarian spokesmen as L. H. Weller, A. P. Hungate, S. O. Daws, Henry Vincent, Charles Macune, W. Scott Morgan, Nelson Dunning, L. L. Polk, S. M. Scott, John B. Rayner, Charles Pierson, Thomas Cater, and Gasper C. Clemens, among others; papers of key opponents of the agrarian organizing drive: A. J. Rose, "Pitchfork" Ben Tillman, and James Hogg; manuscript collections bearing on silver lobbying: William Jennings Bryan, William Allen, Marion Butler, Ignatius Donnelly, Davis Waite, and William Stewart; and, lastly, national, state, and local organization records of the National Farmers Alliance and Industrial Union. The response of the larger society to the farmers' movement was visible in the nation's metropolitan press. These sources, along with rural weeklies, unavoidably create a completely restructured picture of the democratic dynamics of Alliance-Populism. Such a restructuring inevitably leads to a drastic reappraisal of the secondary literature of the agrarian revolt.

Inasmuch as John Hicks's description of the movement has been taken as a general guide to what happened, his pioneering work, *The Populist Revolt,* has strongly influenced all subsequent interpretations irrespective of point of view. Unfortunately, since the Alliance cooperative movement was not seen by Hicks as the core experience of the agrarian revolt, his lengthy work on the shadow movement of free silver has had a crippling influence on subsequent scholarship.

The idea of the shadow movement as the crux of the agrarian revolt governs both the most influential attack on Populism, Richard Hofstadter's *The Age of Reform,* and its most ardent defense, Norman Pollack's *The Populist Response to Industrial America.* Of the two, Hofstadter's study has been far more pervasive in its impact. Correctly finding the free silver arguments of William Jennings Bryan and "Coin" Harvey to be superficial, Hofstadter persuasively indicts what he takes, on Hicksian terms, to be "Populism." He fortifies his analysis through the creation of an elaborately crafted cultural category, which he styled "The Agrarian Myth." Through this device, Hofstadter imputes to Populists a number of modes of self-analysis and national political analysis that were wholly alien to the actual interpretations American farmers achieved as a result of their cooperative struggle. While Hofstadter's misreading has a quality of grandeur, the source of his difficulty is not hard to locate: he managed to frame his interpretation of the intellectual content of Populism without recourse to a single reference to the planks of the Omaha Platform of the People's Party or to any of the economic, political, or cultural experiences that led to the creation of those goals. Indeed, there is no indication in his text that he was aware of these experiences.

Populists are not "intransigents," as they are to Hicks; to Hofstadter, they are scarcely present at all. A Populist indictment of corporate concentration, for example, "reads like a Jacksonian polemic." The dismay of Populists at the practices of American railroad land syndicates and English and Scottish land companies reveals the "anti-foreign" proclivities of people who viewed themselves as "innocent pastoral victims of a conspiracy hatched in the distance." Given the large-scale centralization of land ownership in America in the decades immediately following the agrarian revolt, one is persuaded that the land problems besetting Populists originated a bit closer to home. Similarly, agrarian reservations about the practices of "the town clique"—doubts arising from banker and merchant hostility to the cooperative movement—are seen as providing evidence for Hofstadter's conclusion that Populists suffered from misplaced "anti-urban" manias. For venturesome and creative students of intellectual history such as Richard Hofstadter, the shadow movement of free silver provided a shaky perspective, indeed, from which to interpret Populism.

Similarly, Norman Pollack begins from premises laid down by Hicks and defends the silver crusade as "the last assertion of Populist radicalism." Since Populists were appalled by this particular assertion, Pollack cannot conveniently focus upon their understanding of agrarian purposes. And he does not. While Hofstadter attacks Populism by emphasizing the ephemeral politics of William Jennings Bryan, Pollack defends the movement by defending the same man. Though Hofstadter easily has the best of this discussion, it seems prudent to remember that Bryan was not a Populist. As well as any other, this fact may suggest how far afield the whole matter has been carried. The cooperative dynamics that shaped the Omaha Platform and created the democratic ethos of Populism are not organic to either study.

There is, however, a noteworthy difference in the cultural implications each author draws from the agrarian revolt. On those occasions when Pollack focuses solely on the ideas of Populism, as distinct from specific political and cultural developments presumed to be associated with those ideas, his analysis of the constructive and egalitarian nature of the third party crusade, including its greenback premise, seems unarguable. At such times in *The Populist Response to Industrial America* the shadow movement recedes into the background and Populism emerges. If, on the other hand, in all pertinent interpretive passages concerning "Populism" in *The Age of Reform,* one substitutes the words "proponents of a metallic currency" for the word "Populist," the critique drawn by Richard Hofstadter of the provincialism of Gilded Age America becomes much more precise and persuasive. Such a substitution, of course, alters the meaning of the book at the level of its premise, for *The Age of Reform* then becomes a sweeping criticism of the culture of the corporate state rather than a sweeping criticism of Populism. The democratic rationality of both major parties would then necessarily fall prey to Hofstadterian skepticism.

The narrow limits of Populism that materialize from the work of Hicks, Hofstadter, and Pollack have inevitably exerted a constraining influence on the scores of studies that have subsequently materialized on various aspects of the agrarian crusade. Because the centrality of the cooperative movement has not been understood, historians have found other causes for the presence or absence of Populism in the various regions of America. Where strong third parties emerged, the cause has most frequently been seen to have been "hard times." Where they did not, the weakness of reform was traceable to diversified farming leading to "good harvests" or, conversely, to temporary improvements in commodity prices resulting from "poor harvests." Populism has simply not been seen as a political movement containing its own evolving democratic organization, capable of constructing a mass schoolroom of ideology and a mass culture of self-respect.

Historians of socialist persuasion have encountered similar problems. *The Populist Movement in the United States* by Anna Rochester (1943) broadly follows Hicks's description of the events of the agrarian revolt, as does Matthew Josephson, who analyzed Populism in the course of his broad study of the Gilded Age, *The Politicos* (1938). In at least one instance, a sophisticated and tightly focused local perspective has largely suceeded in skirting Hicksian conceptual limitations: *American Radicalism,* 1865–1901 (1946), Chester McArthur Destler's colorful portrait of the Populist-Socialist alliance in Chicago in 1894. It is the finest local treatment of the movement in all of Populist literature.

Like their socialist counterparts, capitalist historians have tracked the findings of John Hicks, sometimes embellishing their accounts of the shadow movement with Hofstadterian flourishes and sometimes defending the silver episode in the style of Pollack. Stanley B. Parsons subjects the Nebraska silverites to quantification techniques in a recent work entitled *The Populist Context* (1973). In *From Populism to Progressivism in Alabama* (1969), Sheldom Hackney imposes upon rather imprecise and sometimes almost opaque legislative documents a heavy burden of ideological and political interpretation in the course of reaching conclusions in harmony with Hofstadter's. *The Climax of Populism* by Robert F. Durden (1966) is a sympathetic treatment of Marion Butler's harried tenure as third party custodian of the campaign for silver in 1896. Those of his conclusions that concern the efficacy of the silver drive accordingly coincide with Pollack's. In *Populist Vanguard* (1975), a study that emphasizes religious influences more than monetary analysis, Robert McMath finds a significant ingredient of agrarian organizing to be the "congenial social settings" in which the farmers met. In *Farmer Movements in the South* (1960), Theodore Saloutos intermittently observes the cooperative movement of the Alliance but does not see its defeat as a central event. Rather, the Farmers Alliance is "undermined" by the People's Party.

Because the political basis of the national Populist movement has proven such a universal pitfall, several of the most interesting and substantial studies of Populism have been produced by cultural historians who were not primarily dealing with the third party's structural and political evolution. In relating what Populists said to what they did, such writers as O. Eugene Clanton and Walter T. K. Nugent, for example, have provided a much clearer picture of the third party movement in Kansas. Clanton's *Kansas Populism: Ideas and Men* (1969) concludes that Populists were broadly progressive. Nugent, in his more sharply focused study, *The Tolerant Populists* (1963), investigates the day-to-day realities in the life of the third party in Kansas, tests all of Hofstadter's major findings, and finds them inapplicable. Nugent was doubtless aided in his inquiry by his familiarity with monetary issues. His study of post-Civil War financial struggles, *Money and*

American Society, 1865–1880 (1967), probes a number of the differences between greenbackers, silverites, and goldbugs. Robert Sharkey, is even more authoritative on financial matters in *Money, Class, and Party* (1959), but Irvin Unger's *The Greenback Era* (1964), proceeding from Hofstadterian premises, is much narrower and less useful in its description of greenback doctrines. Allan Weinstein's *Prelude to Populism* (1970) is a careful study of the first silver drive of the 1870s, but the author unfortunately relates money to late nineteenth-century politics in ways reflecting the influence of John Hicks.

Five worthy biographies of Populist leaders are now available: C. Vann Woodward's *Tom Watson: Agrarian Rebel* (2nd edition, 1973); Chester McArthur Destler's *Henry Demarest Lloyd and the Empire of Reform* (1963); Martin Ridge's *Ignatius Donnelly: The Portrait of a Politician* (1962); Stuart Noblin's *Leonidas Lafayette Polk: Agrarian Crusader* (1949), and Michael J. Brodhead's *Persevering Populist: The Life of Frank Doster* (1969).

Gilded Age society, marking as it does the beginnings of modern corporate America, offers a fertile field for research on the structure of power and privilege in America. Among the higher priorities, the economic, political, and social ramifications of American banking practices, in both the nineteenth and twentieth centuries, stand as a singularly neglected area of investigation, one affecting hundreds of millions of Americans, past and present. The long-standard works on the origins and development of the Federal Reserve System have emanated from the pens of gold-standard apostles who wrote the Federal Reserve Act—J. Laurence Laughlin, H. Parker Willis, and Paul Warburg. Similarly, the most comprehensive studies of American banking in historical literature are by Bray Hammond, an employee of "The Fed." This material can scarcely be said to constitute a probing or balanced body of evidence. *A Monetary History of the United States*, 1867–1960 by economists Milton Friedman and Anna Schwartz (1963) approaches the politics of money with extreme circumspection, leaving many central issues untouched. *Money* by John Kenneth Galbraith (1975) is urbanely skeptical of a number of sanctioned assumptions and institutions—from the values and intelligence of American bankers to the economic utility of the Federal Reserve System—but the author makes no sustained effort to formulate broadly applicable alternatives. The subjects of money and banking, in their meaning as social and political realities as well as arcane financial topics, do not seem to be in immediate danger of being over-worked by American scholars. Since Populism, serious and full-scale appraisals of the monetary system have not been attempted. A beginning, however, has been made with respect to the helplessness of Congress. *The Money Committees* by Lester Salamon (1974) probes the adverse impact of the banking system and banker lobbying on, among other things, the housing aspirations of millions of Americans.

One also senses that women played a more prominent role in the agrarian revolt than the present study suggests. The evidence is both tantalizing in implication and difficult to gather. Suggested points of entry: the careers of Annie Diggs of Kansas, Sophronia Lewelling of Oregon, Bettie Gay of Texas, Luna Kellie of Nebraska, Ella Knowles of Montana, Sophia Harden of South Dakota, and of course, the author of the famous injunction to farmers to "raise less corn and more hell," Mary Elizabeth Lease of Kansas. Luna Kellie's epitaph for Populism suggests the depth of her personal involvement: "I dared not even think of all the hopes we used to have and their bitter ending . . . and so I never vote."

Of the existing monographic literature, the best single state study of Populism remains A. M. Arnett's *The Populist Movement in Georgia* (1922), written before later interpretive constraints were fashioned. Four recent studies are also of merit. In *One-Gallused Rebellion: Agrarianism in Alabama, 1865–1896,* by William Warren Rogers (1970), a work based on extensive use of primary sources, Alliance-Populism emerges as authentic human striving; and in *Bourbonism and Agrarian Protest* by William Ivy Hair (1969), the third party's checkered struggle against the inher-itance of white supremacy in Louisiana is delineated. *Urban Populism and Free Silver in Montana,* by Thomas Clinch (1970), is a careful study in a state where Populism was a labor movement. The grab-bag nature of the silver crusade is clearly visible in Mary Ellen Glass's *Silver and Politics in Nevada: 1892–1902* (1969).

An older study by a political scientist, Roscoe Martin's *The People's Party in Texas* (1933), is also useful. *Populism and Politics*, by Peter H. Argesinger (1974) is excellent on the machinations of the fusionists in Kansas but less so when the author, basing his work primarily on Kansas sources, attempts to interpret the movement beyond that state's borders.

Over all, the best guides to the national Populist experience remain the Populists themselves. Two thoughtful collections of agrarian thought are available: *The Populist Mind* by Norman Pollack (1967), and *A Populist Reader* by George Tindall (1966). Also useful is "The Rhetoric of Southern Populists: Metaphor and Imagery in the Language of Reform," by Bruce Palmer, a recently completed and as yet unpublished dissertation. For those who want more of the Populists in unvarnished form, the depression year of 1894 brought forth Henry Demarest Lloyd's anti-monopoly classic, *Wealth Against Commonwealth;* Henry Vincent's deeply sympathetic account of the plight of the unemployed, *The Story of the Commonweal,* and, for those who can take a combination of full-blown Populist rhetoric blended with a thoughtful investigation of the monetary system, there is James H. "Cyclone" Davis's *A Political Revelation,* containing a lengthy supplement on the Sub-Treasury System by Harry Tracy.

Some excellent works by specialists bear directly on the issues of Populism. Hans Birger Thorelli explores the rise of oligopoly in his monumental study, *The Federal Anti-trust Policy: Organization of an American Tradition* (1955). With the essential legal flanks safeguarded, the movement toward corporate concentration continued unimpeded into the modern era, as Ralph L. Nelson shows in *The Merger Movement in American Industry, 1895–1956* (1957). While the Justice Department was losing its battle to cope with the combination movement, the Interstate Commerce Commission met a corresponding fate at the hands of American railroads, a process detailed in Gabriel Kolko's *Railroads and Regulation, 1877–1916* (1965). *Triumph of Conservatism* (1963), by the same author, traces the similar accommodations to large-scale manufacturers, merchandisers, and processors by the Federal Trade Commission and to Eastern commercial banks by the Federal Reserve System. The social impact of the emerging corporate state is brilliantly interpreted in *Work, Culture, and Society in Industrializing America,* by Herbert Gutman (1975).

In the presence of consolidating corporate power, the fragility of the American labor movement is visible in Norman J. Ware's old but still useful study of the Knights of Labor, *The Labor Movement in the United States,* 1860–1895 (1929). The analogous plight of Southern farmers under the crop lien system is effectively portrayed by Harold Woodman in *King Cotton and His Retainers: Financing and Marketing the Cotton Crop of the South,* 1800–1925 (1968). However, in a work that treats the problem of agricultural credit in the Western granary, *Money at Interest* (1955), Allan Bogue engages in a somewhat strained defense of the policies of mortgage companies. The matter of tenancy and land centralization in the plains states in both the nineteenth and twentieth centuries can bear some more attention. A relevant body of evidence is available in Fred A. Shannon, "The Status of the Midwestern Farmer in 1900," *Mississippi Valley Historical Review* (December 1950), 491–510. In a recent study, *The Shadow of Slavery: Peonage in the South, 1901–1969* (1972), Pete Daniel has provided a valuable example of what can be achieved on a most relevant and long-neglected subject. The conservative rationale for rural poverty—that farmers brought their troubles on themselves through "overproduction"—is effectively debunked by two economists, an economic historian and a historian, respectively: Roger Ransom and Richard Sutch, *One Kind of Freedom: The Economic Consequence of Emancipation* (1977), which reveals the harshness of the crop lien system; Stephen DeCanio, "Cotton 'Overproduction' in Late Nineteenth Century Southern Agriculture," *Journal of Economic History* (Sept. 1973), 608–33; and Thomas D. Clark, "The Furnishing and Supply System in Southern Agriculture since 1865," *Journal of Southern History* (Feb. 1946), 24–44. The historical record of the cooperative idea is presented in *The Rise of American Cooperative Enterprise* (1969) by Joseph Knapp.

The passing of the People's Party in America left the way clear in the South for the achievement of political hegemony by businessmen. In *The Shaping of Southern Politics* (1974), an effectively documented work that probes the process by which a

white supremacist, business-dominated "Solid South" was constructed, J. Morgan Kousser traces the near-total disfranchisement of blacks and the partial disfranchisement of low income
whites. The new edifice of politics was fully completed by 1910.
The same forces were at work nationally, a process that is often
visible in *The American Party Systems*, edited by William N.
Chambers and Walter Dean Burnham. Should any future historian harbor Hofstadterian doubts that the grievances of Populists and the objectives of the Omaha Platform were real, the
work of these specialists may give him pause.

Finally, the unique contribution of C. Vann Woodward merits
special mention. While concerned with a larger topic, Woodward's classic study of the post-Reconstruction South, *Origins of
the New South* (1951), is laden with insights about the political
evolution and cultural implications of Southern Populism. The
chief limitation of Woodward's analysis is traceable to the
regional scope that was a product of his larger purpose—which
was not to write about the agrarian revolt in the nation but to
write about the American South in the Gilded Age. Woodward's
essential cultural statement about Southern Populism, that it was
demonstrably more humanistic than its political rivals, applies
to American Populism. His essential political statement about
Southern Populists, that they were "mid-roaders," also, of course,
applies to the national movement. Despite a number of impediments in the secondary literature relating to the origins of the
cooperative movement and the evolution of Alliance monetary
theory, Woodward, while not locating the centrality of cooperation or the movement culture it created, drew attention to the
Alliance lecturing system and the National Reform Press Association developed by agrarian strategists as necessary educational
corollaries of their organizing campaign. Accordingly, he had
little difficulty in outlining the structure of Populist economic
reform and the corresponding ephemerality of the silver issue.
In short, in economic as well as political terms, Woodward wrote
about Populism as Populism, rather than as the fusion politics
of free silver. And, of course, Woodward's magnificent biography
of the tortured life of Tom Watson is one of the enduring
triumphs of American historical literature and, indeed, of
American letters.

Index